Probability

Other titles in this series

Already published

Linear Algebra
R B J T Allenby

Numbers, Sequences and Series
K E Hirst

Groups
C R Jordan and D A Jordan

In preparation

Mathematical Modelling
J Berry and K Houston

Discrete Mathematics
A Chetwynd and P Diggle

Particle Mechanics
C Collinson and D Roper

Ordinary Differential Equations
W Cox

Vectors in 2 or 3 Dimensions
A E Hirst

Analysis
E Kopp

Statistics
A Mayer and A M Sykes

Modular Mathematics Series

Probability

John H McColl

Department of Statistics
University of Glasgow

A member of the Hodder Headline Group
LONDON • SYDNEY • AUCKLAND

To Isabel, Richard and Stephanie

First published in Great Britain 1995 by
Edward Arnold
Second impression 1997 by Arnold,
a member of the Hodder Headline Group,
338 Euston Road, London NW1 3BH

British Library of Cataloguing in Publication Data
A catalogue record for this book is available from the British Library

ISBN 0 340 61426 9

Typeset in 10/12 Times by
Paston Press Ltd, Loddon, Norfolk
Printed and Bound in Great Britain by
J W Arrowsmith Ltd, Bristol

Contents

Series Preface

This series is designed particularly, but not exclusively, for students reading degree programmes based on semester-long modules. Each text will cover the essential core of an area of mathematics and lay the foundation for further study in that area. Some texts may include more material than can be comfortably covered in a single module, the intention there being that the topics to be studied can be selected to meet the needs of the student. Historical contexts, real life situations, and linkages with other areas of mathematics and more advanced topics are included. Traditional worked examples and exercises are augmented by more open-ended exercises and tutorial problems suitable for group work or self-study. Where appropriate, the use of computer packages is encouraged. The first level texts assume only the A-level core curriculum.

Professor Chris D. Collinson
Dr Johnston Anderson
Mr Peter Holmes

Preface

A schoolteacher once advised me not to read prefaces, forewords or authors' introductions to their books. I have always followed his advice, so I feel obliged at least to make this preface brief.

Many students complain that probability is 'boring'. Personally, I have always found it fascinating and I hope, in this book, to share my enthusiasm for this subject. Much of my interest in probability is due to the way it can be applied in widely differing subject areas. A number of these are discussed at length in special Applications sections. In addition, many of the examples and exercises are based on real-life problems. I have not totally rejected traditional examples based on coins, dice, etc., since these are useful representations of more complex experiments. For example, the analogy of throwing balls into boxes is used to help explain some results in quantum theory.

Much of this book is based on lectures I give to a second-year class here in the University of Glasgow. It is intended to be suitable as a textbook to accompany courses in the first and second years at a university or college in Scotland and (primarily) in the first year at other universities in the UK. The reader needs a basic knowledge of set theory, though this topic is reviewed briefly in Section 2.4. The book makes use of simple differentiation and integration, though double integration and partial differentiation are avoided since bivariate continuous distributions are not dealt with in detail.

Teachers preparing students for Sixth Year Studies Mathematics (especially Paper III) and A-level statistics might also consider using this book. Sections dealing with the moment-generating function, which might cause particular difficulties at that level, are clearly marked, and the remainder of the book does not depend on them.

Finally, I would like to record my sincere thanks to Mr Peter Holmes for his encouragement and many helpful comments on the manuscript and to staff at Edward Arnold for their help and advice. I would also like to thank my colleagues here at the University of Glasgow for many interesting discussions which have influenced the way in which probability is presented in this book. Lastly, I must thank all of the above, and especially my family, for their patience as this book slowly took shape.

John H. McColl
Department of Statistics
University of Glasgow

1 • Modelling Uncertainty

Why are you studying probability? Do you want to, or do you have to do it as a course requirement? This opening chapter attempts to persuade you that probability is an interesting subject. It first catalogues a variety of practical contexts in which probability is being used. Then it sets probability within the general framework of mathematical modelling.

1.1 Applications of probability

We are all used to uncertainty, and its consequences! Will it rain today? I do not know. Perhaps I should carry my umbrella with me. What score will I get with this die on my next turn? If it is a 6, then I will win the game. Will my house be burgled this year? Hopefully not, but I have insured it just in case.

Uncertainty is part of our daily experience. Newspaper reports commonly use words like 'possibly', 'probably' and 'chance'. We are used to describing events as 'likely' or 'odds on'. Football pools, bookmakers, casinos and the National Lottery testify to human fascination with uncertainty and its consequences.

Probability is the branch of mathematics that seeks to study uncertainty in a systematic way. In the last 200 years, its techniques have been applied successfully in many different fields of study to bring fresh and sometimes surprising insights.

(1) The quantum theory of physics depicts a Universe whose fundamental structure is random. The organization of subatomic particles, atoms in crystals and gas molecules are best described using probabilities.
(2) Modern models of inheritance, too, are based on a probabilistic understanding of how genes are transmitted from parents to children. Genetic counselling relies on this work.
(3) Probability arguments underpin recent modelling of the AIDS epidemic. This is just one example of the many ways in which probability has affected modern medical research and influenced medical decision-making.
(4) Have you ever applied for credit? Your failure or success might well have been due to the calculated probability that you were a good risk. Elsewhere in the financial sector, actuaries have, for generations, used probability models to evaluate life assurance and pension funds, and to set fair levels of premiums and contributions.
(5) Court cases are increasingly being influenced by forensic evidence based on probabilities.

This book will introduce some of these (and other) applications of probability as it goes along, a few in greater detail in special 'Applications' sections. Its main purpose, though, is to present probability as a mathematical discipline that provides a useful framework and specific techniques for studying uncertainty in any context.

1.2 Stochastic experiments

A **probability** is a number that expresses the degree to which an occurrence is certain or uncertain. Conventionally, probabilities are measured on a scale from 0 to 1. Something that is almost certain to happen has a probability close to 1, while an event that is extremely unlikely has a probability close to 0. Most possible happenings have intermediate probabilities. For example, the probability that a coin lands with heads uppermost is often assumed to be $\frac{1}{2}$.

The subject of **probability** is the branch of mathematics that is concerned with building models to describe conditions of uncertainty and providing tools to draw logical consequences on the basis of these models. Fundamental to a meaningful definition of probability is the idea of an **experiment**. In probability, an experiment is basically any process or phenomenon that gives rise to information.

Example 1

Here are brief descriptions of some experiments.

(1) Record the sex of the next baby born at a local maternity hospital.
(2) Interview 100 shoppers in a busy shopping centre; record the number who have heard of a new brand of washing powder.
(3) Test electrical components consecutively as they come off a production line; record the number tested until the first defective component is discovered.
(4) Grow a new strain of tomato plant; record the yield of tomatoes (kilograms) from the plant.
(5) Operate on a patient to remove a stomach cancer; record the amount of time (months) that goes by until the patient relapses.
(6) Record the birthweights (kilograms) of twins.

Notice that only some of these are experiments in the usual scientific sense. Though the experimenter has control over some of the experimental conditions in (3), (4) and (5), in (1), (2) and (6) the experimenter is just a passive observer.

The information that is recorded as a result of an experiment is called the **outcome** of the experiment. An outcome need not be numerical; for example, in experiment (1) above, the information recorded is whether the child is a girl or a boy. An outcome may consist of more than one item of information; for example, in experiment (6), the outcome consists of the separate birthweights of the two children.

All of the examples above are **stochastic (or random) experiments**. In each case, there is more than one possible outcome of the experiment, and the one that will actually occur cannot be known with certainty before conducting the experiment. For example, in experiment (1), the first baby born might be either a boy or a girl; we will not know for certain until it is born. If we repeat the experiment one hour later in the same hospital, then we might observe a different outcome. In experiment (2), either 0 or 1 or 2 or ... or 100 shoppers might have heard of the washing powder; we will not know exactly how many until we ask them. If we select a different sample of 100 shoppers, even in the same shopping centre on the same day, then we will generally observe a different outcome.

Experiments that are not stochastic are called **deterministic**, meaning that the experimental conditions uniquely determine the outcome. For example, suppose that we connect a battery (V volts) into a simple circuit (resistance R ohms). The resulting current (in amperes) is $A = V/R$, by Ohm's Law. No matter how often we carry out this experiment, as long as we hold the experimental conditions (V and R) constant then we will obtain the same outcome A. Admittedly, with very accurate measurement, we might be able to detect slight fluctuations in A from experiment to experiment, due to so-called experimental error, but these would be negligible in comparison with A itself. The experimental conditions have determined A uniquely (to all intents and purposes).

Probability is the mathematical study of stochastic (rather than deterministic) experiments. A single performance of a stochastic experiment is called a **replicate**. Some probabilists restrict probability to the study of stochastic experiments that (at least conceptually) may be replicated a large number of times under identical experimental conditions. They might, therefore, deny any mathematical meaning to the phrase 'the present government has probability 0.2 of winning the next General Election', on the grounds that the next election will happen once only and is not repeatable under identical conditions. We shall briefly discuss this matter again in Chapter 3.

1.3 Mathematical models

This book presents probability within the general context of mathematical modelling. A probability model is an abstract, mathematical representation of a stochastic experiment. By its nature, a probability model cannot capture every detail of an experiment, for then it would be as difficult to analyse as the reality it seeks to represent. To be of practical use, though, the model must represent faithfully those features of the experiment that are important in determining its outcome. Figure 1.1 portrays the process of mathematical modelling.

The process of modelling a stochastic experiment begins in the real world. We have some information about the background to the experiment; perhaps we know the outcomes from previous replicates of it or from replicates of similar

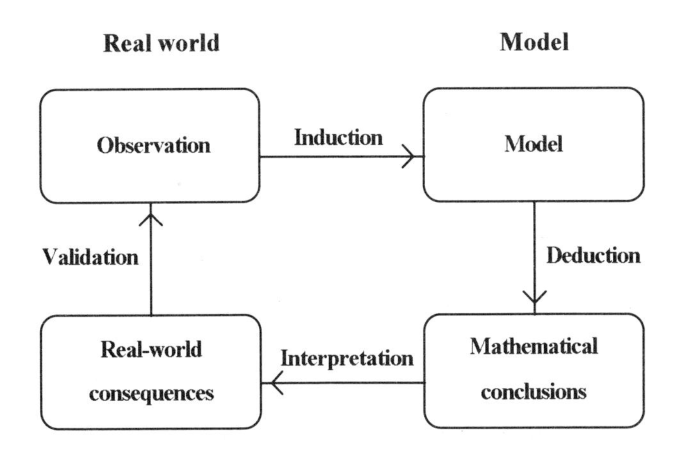

Fig 1.1 The process of mathematical modelling.

experiments. On the basis of all that we know, we construct a probability model. This is not a matter of precise mathematical deduction; it is an **inductive**, usually intuitive, step, which builds on our own and others' experience within the specific application area as well as our general knowledge of probability models. This book introduces some of the standard probability models that have been found to be useful in many different contexts. It also points out the experimental features to look for to justify the adoption of these models in any particular case.

The next step is the most narrowly mathematical. By making logical **deductions** within the model, we derive mathematical conclusions. The name probability calculus is often given to the system of axioms, rules and techniques that governs this part of the process. Much of this book is devoted to introducing this theory (the foundations being laid in Chapters 5 to 7).

We cannot be satisfied by deriving conclusions within the model itself. Our third step is to **interpret** our mathematical conclusions in terms of their real-world consequences. In all the examples in this book, care is taken to indicate the practical meaning of any calculations we carry out.

Fourthly, we test or **validate** our model. On the basis of data (new or old) we decide whether the real-world behaviour predicted from the model is in good agreement with what we actually observe. If it is, then we are content with our model. Validation is, properly, the concern of statistical inference rather than probability, and we will not refer to it again in this book.

More often than not, we will be unhappy with some aspect of our model. We can usually think of some way to modify it, to make it more realistic. So we **iterate** through the whole modelling process again and again until we eventually find an acceptable model.

Chapter 2 describes the first steps we must take in building any probability model.

Summary

This chapter has tried to show that probability is interesting and worth studying because of its widespread use in all kinds of areas. Any phenomenon or process that yields information is called an experiment. The information itself is called the outcome of the experiment. A stochastic experiment is distinguished by having a number of different possible outcomes for the same experimental conditions. Only one outcome will occur on any particular occasion, and which one it will be is unknown beforehand. Probability is the mathematical study of stochastic experiments. It is concerned with building, and making logical deductions within, consistent models for stochastic experiments. Successful models give real-world predictions that accurately mimic observed behaviour.

2 • Sample Spaces and Events

This chapter introduces the most basic feature of a probability model, the sample space, which is a list of all the possible outcomes of a stochastic experiment. It also uses set theory to show how events associated with the outcome of an experiment may be represented within the sample space.

2.1 The sample space

A stochastic experiment is a phenomenon or process with a number of different possible outcomes. Whenever the experiment is performed, one and only one of the possible outcomes will occur, but which one cannot be known with certainty beforehand.

The collection of all possible outcomes is called the **sample space** for the experiment. Mathematically, it is convenient to describe the sample space as a set. (Set theory is reviewed briefly in Section 2.4, for any reader who would like a reminder of the basic concepts.)

Formally, then, a **sample space**, S, for a stochastic experiment is a set in which every outcome of the experiment is represented by one and only one element.

Example 1

Record the number of defective light bulbs in a box of ten light bulbs. A suitable sample space is $S = \{0, 1, 2, 3, 4, 5, 6, 7, 8, 9, 10\}$.

Example 2

Count the number of days this week on which it rains. A suitable sample space is $S = \{0, 1, 2, 3, 4, 5, 6, 7\}$.

Example 3

Count the number of cars passing a fixed point on a motorway in one minute. A suitable sample space is $S = \{0, 1, 2, 3, \ldots\}$.

This last set might seem too big to be a sensible sample space for this experiment. No doubt every possible outcome of the experiment is represented by a unique element in S. But what about elements of S such as 1,000,000? It is surely impossible for 1 million cars to pass a fixed point in one minute. This objection to S can be answered, first, by pointing out that it would be very difficult to specify a cut-off point so that S consisted only of plausible outcomes. Is 1000 cars a possible outcome but 1001 cars impossible? Secondly, it is mathematically convenient to use the set of whole numbers as the sample space. We are usually satisfied with a sample space which is at least big enough to represent every plausible outcome.

Notice that every possible outcome, however unlikely, must be represented by an element in the sample space.

Example 4

Record the height (metres) of a tree in a forest. A suitable sample space is $S = \{x : x > 0\}$.

Example 5

Record the birthweights (kilograms) of twins. A suitable sample space is $S = \{(x, y) : x > 0 \text{ and } y > 0\}$.

Example 6

Often, we have a choice of quite different sample spaces for an experiment. Suppose that we want to record the sexes of the children in a family with three children. The most informative sample space is

$$S = \{\text{fff, ffm, fmf, fmm, mff, mfm, mmf, mmm}\},$$

where, for example, ffm represents the (ordered) outcome that the two oldest children are female and the youngest child is male.

If only the total numbers of boys and girls in the family were important to us, and not their birth order, we could obtain all the information we want by recording just the number of girls in the family. Implicitly, we would then be adopting the alternative, less informative, sample space:

$$S_1 = \{0, 1, 2, 3\}.$$

EXERCISE ON 2.1

1. Write down suitable sample spaces for recording:
   (a) the number of games of badminton you win in a series of three games with a friend;
   (b) the number of times you attend the doctor in a calendar year;
   (c) the length of time (minutes) it takes a Fire Brigade to respond to an emergency telephone call;
   (d) the difference in the heights (metres) of a husband and wife;
   (e) the length of time (minutes) you have to wait at the bank before being served;
   (f) the length of time (minutes) you have to wait at the bank before being served, and the length of time (minutes) the teller subsequently takes to deal with your business;
   (g) the number of correct answers given in a general knowledge quiz by a contestant who is asked 100 questions;
   (h) the numbers of correct answers given by each of two contestants in a general knowledge quiz, where each separately is asked 100 questions.

2.2 Events

An **event** is any occurrence that results from the performance of an experiment. Every outcome is an event, known as an **elementary event**. The term is used more widely to mean any collection of outcomes; an event that consists of more than one outcome is called a **compound event**. Mathematically, an event is any happening that may be represented by a subset of the sample space.

Example 1 (continued)

The elementary event 'one light bulb is defective' is represented by the subset $\{1\}$. The compound event 'at least one light bulb is defective' is represented by the subset

$$E = \{1, 2, 3, 4, 5, 6, 7, 8, 9, 10\}$$

and the event 'at most one light bulb is defective' is represented by the subset

$$F = \{0, 1\}.$$

Clearly, there is a distinction between an event itself and the subset of S that we use to represent it. In the remainder of this book, though, we will neglect this difference and write, for example, E to mean both the event 'at least one light bulb is defective' and the subset $\{1, 2, \ldots, 10\}$.

Example 4 (continued)

The event G = 'the tree is between 5 and 6 metres tall' is represented by the subset $\{x : 5 \leq x \leq 6\}$.

S itself represents any event that is *certain* to occur. The empty set, $\{\ \}$ or $\emptyset$, represents an *impossible* event.

EXERCISES ON 2.2

1. Here are descriptions of events associated with the experiments described in Exercise 1 after Section 2.1. Write out each event as a subset of the corresponding sample space.
   (a) A = 'you win at least two games'; B = 'your friend wins at least two games';
   (b) E = 'you attend the doctor no more than twice';
   (c) F = 'the Fire Brigade responds in less than ten minutes'; G = 'the Fire Brigade takes more than five minutes to respond';
   (d) A = 'the wife is taller than her husband';
   (f) C = 'the time you wait to be served is longer than the time that it takes the teller to serve you';
   (h) D = 'the first contestant gives at least 75 correct answers'; E = 'the second contestant gives at least 75 correct answers'; F = 'between them, the contestants correctly answer at least 150 questions'.
2. You intend to toss a coin three times. Write down sample spaces to record:
   (a) the outcomes of the three tosses, in the order in which they occurred;
   (b) the total number of heads obtained;
   (c) the number of tails obtained before the first head is obtained.

In which of these sample spaces can you represent the following events:
A = 'tails is the result of the first toss';
B = 'at least one head is obtained before the first tail';
C = 'two heads and one tails are obtained';
D = 'the order of the results is heads, then tails, then heads'?

2.3 Relationships among events

Just as events can be represented in a mathematical model by subsets of the sample space S, important relationships among events can be represented by set relationships among the corresponding subsets. It is particularly important to understand what events are represented by the intersection and the union of two (or more) events.

Example 7

Suppose that a doctor has to make three house calls, to patients X, Y and Z, but can choose the order in which he or she will make the calls. There are six possible (ordered) outcomes:

$$S = \{XYZ, XZY, YXZ, YZX, ZXY, ZYX\}$$

where, for example, XYZ means that the doctor visits X first, then Y, then Z. Define the events E = 'the doctor visits X before Y', F = 'the doctor visits Y before Z' and G = 'the doctor visits Z before X'. Then

$$E = \{XYZ, XZY, ZXY\}$$
$$F = \{XYZ, YXZ, YZX\}$$
$$G = \{YZX, ZXY, ZYX\}$$

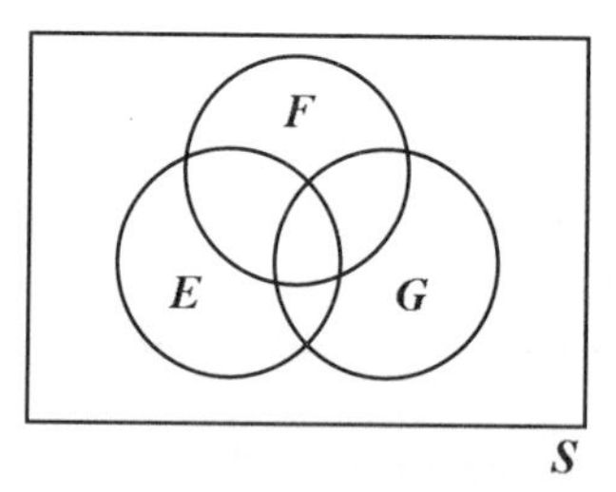

These events are represented in the Venn diagram above, using S as the *universal set*.

The *complement* of E is $E' = \{YXZ, YZX, ZYX\}$, which represents the event 'the doctor visits X after Y', or 'the doctor does *not* visit X before Y'; in other words, E' represents the event 'E does *not* occur'.

The *intersection* of E and F is $E \cap F = \{XYZ\}$, which represents the event 'the doctor *both* visits X before Y *and* visits Y before Z'; in other words, $E \cap F$ represents the event '*both* the event E *and* the event F occur'. In a similar way, $F \cap G = \{YZX\}$ and $G \cap E = \{ZXY\}$.

$E \cap F \cap G = \{\ \}$, *the empty set*: clearly, the event 'the doctor visits X before Y *and* visits Y before Z *and* visits Z before X' is impossible.

The *union* of E and F is $E \cup F = \{XYZ, XZY, YXZ, YZX, ZXY\}$, which represents the event '*either* the doctor visits X before Y *or* the doctor visits Y before Z (*or both*)'; in other words, $E \cup F$ represents the event '*either* the event E occurs *or* the event F occurs (*or both* occur)'.

$E \cup F \cup G = S$ itself: the event 'the doctor *either* visits X before Y *or* visits Y before Z *or* visits Z before X' is certain to occur.

Example 8

A gas central-heating system consists of a pump and a boiler. In the course of a winter both may fail. If either component fails, the system itself fails to operate. Suppose that the system is run for one winter and that an outcome (x, y) is recorded: $x = 0$ if the pump runs throughout the winter without failing; otherwise, $x = 1$. Similarly, $y = 0$ if the boiler runs through the winter without failing; otherwise $y = 1$.

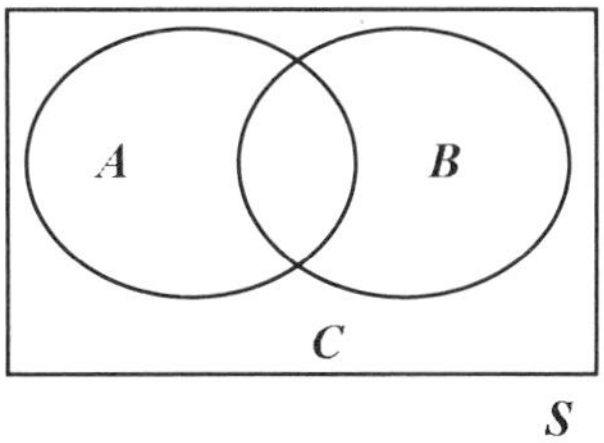

A sample space for this experiment is $S = \{(0,0), (0,1), (1,0), (1,1)\}$. The event A = 'the pump fails at some point during the winter' is represented by $(1,0), (1,1)\}$. The event B = 'the boiler fails at some point during the winter' is represented by $\{(0,1), (1,1)\}$. The event C = 'the system runs throughout the winter' is represented by $\{(0,0)\}$.

$A \cap B = \{(1,1)\}$ represents the event '*both* the pump *and* the boiler fail during the winter'. On the other hand, $A \cup B = \{(0,1), (1,0), (1,1)\}$ represents the event '*at least one* of the components fails during the winter'. Notice that $A \cup B = C'$.

$C \cap A = \{\ \}$, the empty set. It is impossible for the system to run throughout the winter if the pump has broken down. We say that the events C and A are **disjoint** if there are no outcomes in both sets. In this case, C and B are also *disjoint*.

In fact, $A \subseteq C'$: if event A occurs, then event C' must occur; if the pump breaks down, then the system must break down.

Table 2.1 summarizes the uses we have made of set notation to represent relationships among events in a sample space.

Table 2.1 Some uses of set notation to represent relationships among events.

Relationship among events	Set notation
An event that is certain to occur	S
An event that is impossible	$\emptyset$
Event E does not occur	E'
Both the events E and F occur	$E \cap F$
Either E or F occurs or both events occur	$E \cup F$
If E occurs, then F must occur also	$E \subseteq F$
E and F are disjoint (cannot both occur)	$E \cap F = \emptyset$

EXERCISE ON 2.3

1. In the examples of Exercise 1 after Section 2.1 and Exercise 1 after Section 2.2, write out the following subsets of the sample space, and write down a description in words of the events they represent.
   (a) $A \cup B, A \cap B, A'$;
   (b) E';
   (c) $F \cap G, F \cup G$;
   (h) $D \cap E, D \cap E \cap F$.

2.4 A review of set theory

(This section may be omitted, or glanced over quickly, by a reader who is familiar with set theory.)

A **set** is just any well-defined list or collection of objects. For example:

(1) the set of vowels in the English alphabet is $V = \{\text{a, e, i, o, u}\}$;
(2) the set of people who live in my house is $H = \{\text{John, Isabel, Richard, Stephanie}\}$;
(3) the set of digits, $D = \{0, 1, 2, 3, 4, 5, 6, 7, 8, 9\}$.

It is conventional to designate a set by a capital letter. As above, when a set contains a finite number of objects, we use curly brackets, { and }, to denote the start and end of a listing of them all. When the set is infinite, that is it contains infinitely many objects, we must modify this notation. Some infinite sets may be written out in a list that is shown to continue indefinitely, for example:

(4) the set of natural numbers, $\mathbf{N} = \{1, 2, 3, \ldots\}$.

Sets that may be listed in the general form $\{s_1, s_2, s_3, \ldots\}$ are called **countable**. Infinite sets that are related to the set of real numbers, **R**, are usually **uncountable**. For example, it is impossible to write out a list of all the real numbers between 0 and 1. We can use the following notation to define a subset of **R** by some property that all its elements share, for example:

(5) the set of all real numbers between 0 and 1 (inclusive) is $\{x : 0 \leq x \leq 1\}$;
(6) the set of non-negative real numbers, $\mathbf{R}^+ = \{x : x \geq 0\}$.

The individual objects in a set are called its **elements** or **members**. If an object, s, is an element of the set S, we write $s \in S$. If s is not a member of the set S we write $s \notin S$. For example, $7 \in \mathbf{N}$ but $-2 \notin \mathbf{N}$. A special set is the so-called **empty set**, one that has no elements. This can be written { }, or by convention as $\emptyset$.

It does not matter which order the elements of a set are written in. For example, the set of digits, D, could also be written $\{9, 3, 5, 8, 0, 1, 4, 7, 6, 2\}$, though this would be less natural than the usual order shown above.

A set A is called a **subset** of S if every element of A is also an element of S. We write this relationship as $A \subseteq S$. In the examples above,

$$\{\text{a}, \text{e}, \text{u}\} \subseteq V$$
$$\{\text{John}\} \subseteq H$$
$$\mathbf{N} \subseteq \mathbf{R}$$
$$\{x : 0 < x < 1\} \subseteq \mathbf{R}^+$$

Now, any set must be a subset of itself: $S \subseteq S$. It is mathematically convenient to consider the empty set, $\emptyset$, as a subset of every set: $\emptyset \subseteq S$.

Venn diagrams provide a useful method of displaying relationships between subsets of a given set, S. In this context, S is called the **universal set**. S itself is represented by the interior of a rectangle and its subsets by regions within the rectangle, for example the interiors of ellipses or circles. No meaning is attached to the boundaries of the regions.

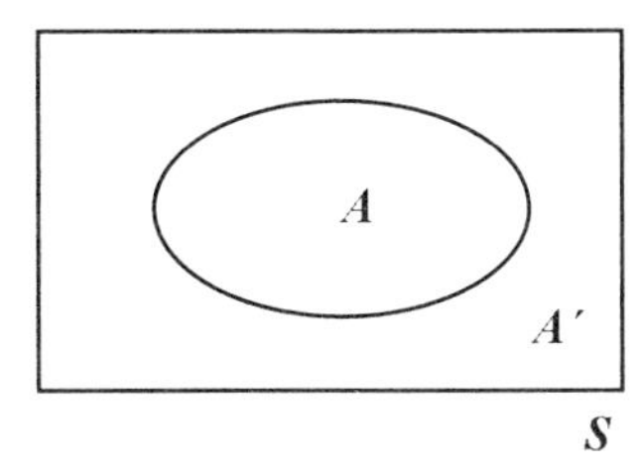

Here is a Venn diagram that shows a single subset $A \subseteq S$. The **complement** of A, written A', is the set of all elements of S that are not members of A. In symbols,

$$s \in A' \Leftrightarrow s \notin A$$

Clearly, $(A')' = A$.

We write $A \subseteq B$, A is a **subset** of B, if every element of A is also a member of B. If $A \subseteq B$ and $B \subseteq A$, then $A = B$.

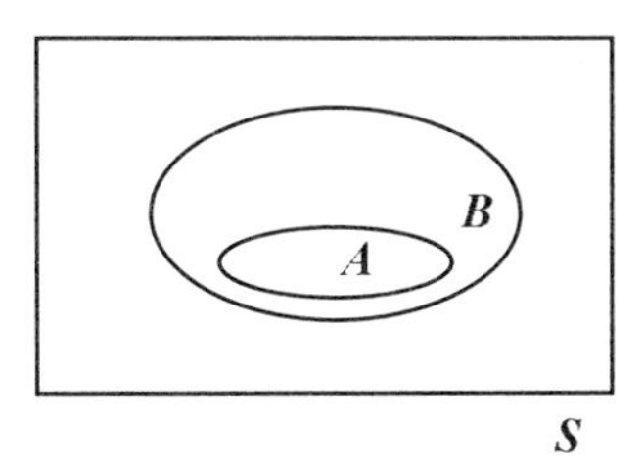

The **intersection** of two subsets A and B, $A \cap B$, is the set of all elements of S that are common to A and B. In symbols,

$$s \in A \cap B \Leftrightarrow s \in A \quad \textit{and} \quad s \in B$$

Clearly, $A \cap B = B \cap A$.

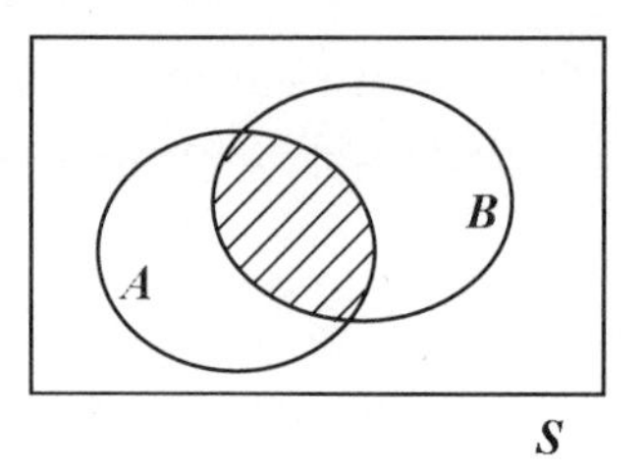

$A \cap B \cap C$ is defined to be the set of all elements of S that are common to A, B and C. This can also be written as $(A \cap B) \cap C$ or $A \cap (B \cap C)$. Clearly,

$$A \cap B \cap C = A \cap C \cap B = B \cap A \cap C = \ldots$$

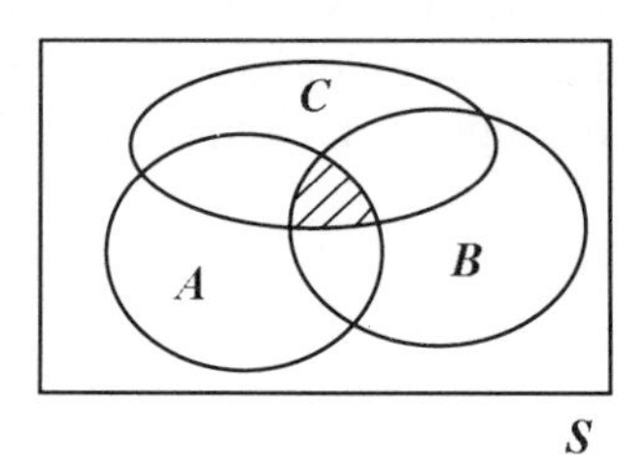

In a similar way, we can define the intersection of any sequence of sets, $A_1, A_2, \ldots$, and we write this set as

$$\bigcap_i A_i$$

The **union**, $A \cup B$, is the subset of S consisting of all objects that are members of A or B or both. In symbols,

$$s \in A \cup B \Leftrightarrow s \in A \quad or \quad s \in B$$

Clearly, $A \cup B = B \cup A$.

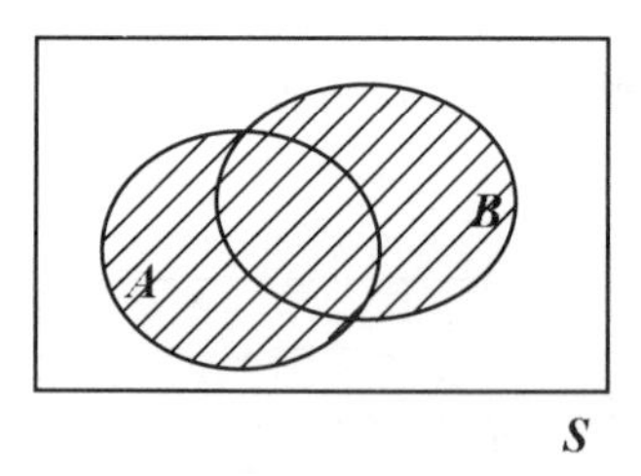

$A \cup B \cup C$ is the subset of S consisting of all objects that are members of at least one of the subsets A, B and C. This can also be written as $(A \cup B) \cup C$ or $A \cup (B \cup C)$. Clearly,

$$A \cup B \cup C = A \cup C \cup B = B \cup A \cup C = \ldots$$

In a similar way, we can define the union of any sequence of sets, $A_1, A_2, \ldots$, and we write this set as

$$\bigcup_i A_i \emptyset$$

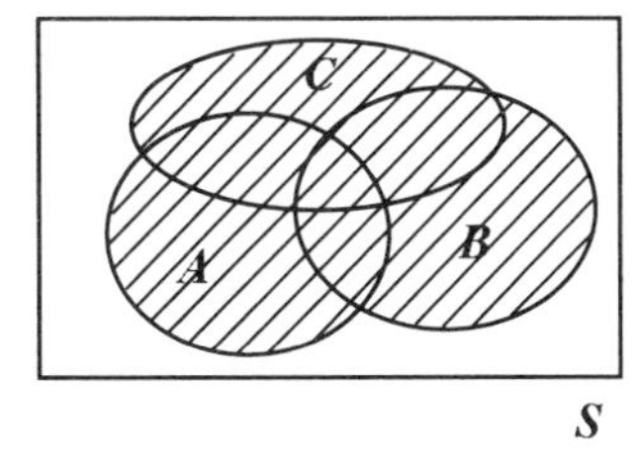

Notice that, though the events $A \cup B \cup C$ and $A \cap B \cap C$ are well defined, $A \cup B \cap C$ is not. This could mean either $(A \cup B) \cap C$ or $A \cup (B \cap C)$, two sets that are not in general equal. Which set is intended must be indicated by using appropriate brackets. The **distributive laws** for sets state that

$$(A \cup B) \cap C = (A \cap C) \cup (B \cap C)$$

$$A \cup (B \cap C) = (A \cup B) \cap (A \cup C)$$

Summary

This chapter has introduced the notion of a sample space as a set in which every possible outcome of an experiment is represented by one and only one element. Any event can be represented by a subset of the sample space (which is the universal set for this purpose). Relationships among events can be expressed within the mathematical model using the set notation of complements, subsets, unions and intersections.

FURTHER EXERCISES

1. Write down a suitable sample space for each of the following experiments.
(a) You record the number of times you win a new computer game in five attempts.
(b) You record the total number of attempts you require in order to win a new computer game for the first time.

2. A standard die is a cube, whose six faces are marked with the numbers $1, 2, \ldots, 6$. The score obtained from rolling a die is the number on the face that lands uppermost. Write down a sample space, S, for the experiment of rolling a standard die once. Write out the following events as subsets of S:
A = 'the score obtained is an even number';
B = 'the score obtained is at most 3';
C = 'the score obtained is 6'.
Then write out the following subsets of S and write down a brief description of the events they represent:
$A', B', A \cup B, A \cup C, B \cup C, (A \cup B \cup C)', A \cap B, A \cap C$

3. A race in an international athletics competition is being contested by three Americans, two Britons, two Canadians and one Dane. Write down a suitable sample space for recording the nationalities of the athletes who come first and second, in that order. Let the events
E = 'one of the Britons wins';

$F =$ 'one of the Americans wins';
$G =$ 'one of the Americans comes second'.
Write out the following subsets of S:
$E, F, G, (E \cup F) \cap G, E \cup (F \cap G), E \cup F \cup G, E \cap F \cap G$

4. Four people get into a lift at the ground floor of a building that has seven higher floors. Write down a suitable sample space for each of the following experiments:
(a) recording the number of different floors at which the passengers get out;
(b) recording the number of passengers who get out at each floor;
(c) recording the floor at which each passenger gets out.
In which of these sample spaces can you represent the following events:
$E =$ 'no passenger gets out at Floor 7';
$F =$ 'all the passengers get out at the same floor';
$G =$ 'the passengers all get out at different floors'?

5. In the Premier Division of the Scottish Football League, teams play each other four times each season. A team gains three points for a win, one point for a draw and no points for a defeat.
(a) Suppose that you are interested in the total numbers of wins for team A, draws, and wins for team B in their games this season. Explain why you need only record two pieces of information, and write down a suitable sample space.
(b) Suppose that you are interested in the total numbers of points scored by teams A and B in their games against each other this season. Explain why you need to record both pieces of information, and write down a suitable sample space.

6. Suppose that E and F are events in a sample space, S.
(a) Suppose that $E \cap F = \emptyset$. Explain why $E \cap F \cap G = \emptyset$, for any $G \subseteq S$.
(b) Suppose that $E \cup F = S$. Explain why $E \cup F \cup G = S$, for any $G \subseteq S$.

3 • Definitions of Probability

In the last chapter, we began the process of modelling a stochastic experiment by defining a sample space to represent all its possible outcomes, and by using subsets of the sample space to represent events associated with the experiment. This chapter makes a formal definition of the probability of an event within a sample space.

3.1 Relative frequency

Having defined the sample space and events within it, the basic building blocks of a mathematical model of a stochastic experiment, we can now proceed to define the probability of an event. From the basic definition, we can then begin to derive procedures for combining known or assumed probabilities to determine the probability of another event. In other words, proceeding from certain assumptions we shall, as in all mathematics, investigate logical consequences.

People think about, and define, the concept of a probability in different ways. As it turns out, this does not affect the way in which probabilities are manipulated mathematically, since all sides agree about the algebraic and arithmetical procedures that must be followed to produce consistent results. Probabilists differ fundamentally, though, in the meaning they attach to their answers.

The view of probability that has been most widely held in recent times is known as the **frequentist** approach. We can introduce it through the following example.

Example 1

A new drug to reduce blood pressure in patients suffering from high blood pressure (hypertension) is being tested on all the new cases referred to a group of hypertension clinics in the course of a year. About 100 new patients are referred each month. The outcome of interest is whether or not each patient's blood pressure is controlled, or brought into a normal range, by a course of treatment. Table 3.1 gives the (simulated) results for 100 patients each month.

For the underlying stochastic experiment of administering the new drug treatment to an individual, a suitable sample space is $S = \{\text{controlled, not controlled}\}$.

In the first month, the basic experiment is repeated (or replicated) 100 times. The event E = 'blood pressure is controlled' occurs on 80 out of the 100 replicates. The **relative frequency** of this event is just the proportion of replicates on which E occurs, in other words 80/100 or 0.8. We can write $rf_{100}(E) = 0.8$.

In the first two months, there are 200 replicates in total, that is at the end of two months, the **cumulative** number of replicates is 200. The cumulative number of replicates on which the event E occurs is 168, so the **cumulative relative frequency** of the event E is 168/200 or 0.84. We can write $rf_{200}(E) = 0.84$. And so on.

Looking down the columns of the table, we can see that there are quite big fluctuations in the monthly success rates, the poorest being 75/100, the best

Table 3.1 Simulated results for Example 1.

Month	Number of subjects	Number controlled	Cumulative number of subjects	Cumulative number controlled	(Cumulative) relative frequency
1	100	80	100	80	0.800
2	100	88	200	168	0.840
3	100	75	300	243	0.810
4	100	77	400	320	0.800
5	100	80	500	400	0.800
6	100	76	600	476	0.793
7	100	82	700	558	0.797
8	100	79	800	637	0.796
9	100	80	900	717	0.797
10	100	76	1000	793	0.793
11	100	77	1100	870	0.791
12	100	78	1200	948	0.790

88/100. However, the cumulative relative frequency in the last column rapidly stabilizes at a value between 0.790 and 0.800, in spite of the fluctuations from month to month. This can be seen even more clearly from a plot of the cumulative relative frequency, $rf_n(E)$, in Fig. 3.1.

It is a recorded feature of stochastic experiments in general that, when they are replicated under identical conditions, then the relative frequencies of events do settle down in this way. There can be no mathematical proof of this statement; it is simply an empirical observation. This phenomenon is known as **statistical regularity**. It is fundamental to the frequentist definition of probability.

Finally, we are in a position to make a formal definition of the probability of an event. The **probability** of the event E, written $P(E)$, is defined to be the long-run relative frequency of E, that is

$$P(E) = \lim_{n \to \infty} rf_n(E)$$

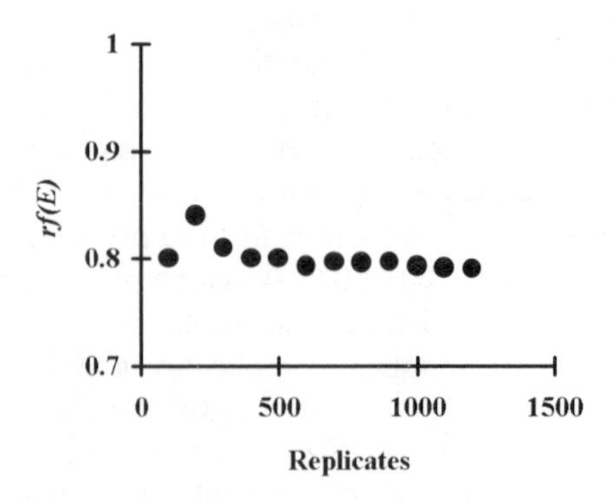

Fig 3.1 Plot of the cumulative relative frequency, $rf(E)$, in Example 1.

In Example 1, on the basis of one year's results (i.e. tests on 1200 patients), we would judge that the probability that this drug controls a hypertensive subject's blood pressure is $P(E) \cong 0.79$.

It might seem from the above discussion that the evaluation of a probability always depends on experimentation. This is not in fact the case. For example, when tossing a 'fair' coin, we believe that the probability of heads landing uppermost is $\frac{1}{2}$. The symmetry of the experiment leads to this belief without experimentation; if we tossed the coin many times, we would expect to get heads on roughly $\frac{1}{2}$ of them. This is the basis of an older (**classical**) approach to probability, which we shall explore in the next chapter.

There are many cases of this kind, where probabilities are assigned to outcomes as a result of reflection on the experimental conditions, without performing the experiment in the first instance. Of course, the resulting probability model would have to be validated from real-world observations, as we discussed in Chapter 1.

EXERCISES ON 3.1

1. The table below gives some information from a recent class experiment. Each member of the class, at some point on the way home one evening, observed a stream of 20 cars and recorded the number of them that were blue. Calculate the cumulative relative frequencies of the event 'car is blue' after each observer's data have been counted in. Plot the cumulative relative frequencies appropriately and show that these data are consistent with statistical regularity. What would you say is the probability that a car in the Greater Glasgow area is coloured blue?

Observer	1	2	3	4	5	6	7	8	9	10	11	12	13	14	15	16
No. of blue cars out of 20 cars	6	5	4	5	5	9	3	4	5	5	4	7	5	6	6	5

2. The members of the same class each tossed a 10p coin 20 times. The results are given in the table below. Do these data seem consistent with the belief that the probability of getting 'heads' when tossing a 10p coin is $\frac{1}{2}$?

Observer	1	2	3	4	5	6	7	8	9	10	11	12	13	14	15	16
No. of 'heads' in 20 tosses	10	9	10	8	8	10	9	7	8	9	9	9	13	10	11	8

3.2 Subjective probabilities

In recent years, the frequentist approach to probability has come under increasing fire from opponents who feel that it restricts the scope of the subject.

Conceptually, the definition given above is tied to stochastic experiments that are capable of replication under identical conditions. It would not be possible on the frequentist approach to assign a probability to an event such as 'the Labour Party will form the next government of the UK', since that is an outcome of an experiment (the next General Election) that is not, by its nature, repeatable.

The 'subjectivist' school attempts to give a meaning to probability that allows such cases to be included in the definition. Their solution is to break the (at least, apparent) link between the frequentist definition of probability and objective reality. A probability, they claim, is a subjective statement about an individual's degree of belief in the chance of an outcome, based on that individual's knowledge of the experimental conditions. The probability of an event, therefore, varies from observer to observer.

How should an individual determine his or her probability of an event? Here most subjectivists would suggest a procedure based on betting odds. If the probability of an event E is $P(E)$, then the probability that E does not occur is $1 - P(E)$. (We will see this later, in Chapters 4 and 5.) The **odds** on E occurring, then, are the ratio of $P(E)$ to $1 - P(E)$. For example, if the probability that a horse will win a race is 0.5, then the odds on the horse winning are 1 to 1, or 'evens'. If the probability that the horse will win is 0.333, then the odds on the horse winning are 1 to 2, or 'two to one against'.

An individual's subjective probability of an event should correspond to the lowest odds in favour of the event at which the individual would be willing to place a bet on that event happening. So, if you think that the odds on the Labour Party forming the next UK government are at least 'evens', and you would be willing to place a bet at these odds, then your subjective probability for this event is 0.5.

For the rest of this book, it will not matter much whether you think of a probability as a long-run relative frequency or as a personal expression of the degree of uncertainty. Though there is disagreement about what a probability means, there is no disagreement about how probabilities must be manipulated to give consistent results. (An excellent, though advanced, discussion of the nature of probability is given in Chapter 3 of Barnett (1982).)

Summary

This chapter has introduced two (competing) definitions of what a probability means. According to the frequentist definition, a probability is a long-run relative frequency, or the proportion of replicates of the experiment on which an event occurs. This is justified, for experiments which can be replicated under identical conditions, by the phenomenon known as statistical regularity. The subjectivist school attempts to widen the definition to include probabilities of 'one-off' events, by insisting that a probability can also represent an individual's degree of belief in the various possible outcomes of an experiment. Subjective probabilities can vary from person to person.

TUTORIAL PROBLEM 1

Investigate the probabilities of events associated with team sports you are interested in, for example the probability that a league football match ends in a draw or that at least one century is scored during a county cricket match. Take the weekly results from a newspaper and plot the cumulative relative frequencies for the whole of a season. Is there convincing evidence of statistical regularity? If not, why not?

TUTORIAL PROBLEM 2

Of all the children born in the UK, about 49% are girls. The aim of this project is to investigate whether parents are as likely to place a newspaper announcement of the birth of a daughter as of the birth of a son. Choose a local newspaper and a national newspaper (such as *The Times*). Each day for a month record the total number of birth announcements and the number that relate to girls. Do not count announcements of twins, triplets, etc. Plot the cumulative relative frequencies daily. To what value does the relative frequency settle down?

TUTORIAL PROBLEM 3

It is claimed that different authors use even common words (such as 'and', 'but', 'a' or 'the') with different frequencies which are characteristic of the individual author's style. This is the basis of research into documents of disputed authorship, for example of the 'Federalist papers' which were influential in the struggle for American Independence, and of the Pastoral Epistles of the New Testament. Take a work or works of an author you enjoy reading and investigate the relative frequencies of some common words (see also Exercise 2 of the Further Exercises below). Choose (say) 20 pages of the book at random (a table of random digits is given in Appendix 1) and identify an excerpt of 100 words on each page. Count the number of occurrences of your chosen word(s) in these passages. Plot the usual cumulative relative frequencies and investigate the statistical regularity.

FURTHER EXERCISES

1. An inspector in a factory that makes light switches took a sample of 50 switches off the production line during each of the ten shifts in a working

week. The inspector tested each of them and recorded the following numbers of defective switches for the ten shifts:

6, 4, 7, 6, 3, 4, 8, 12, 14, 9.

Plot the cumulative relative frequency of the event 'a switch is defective' at the end of each shift. Why do you think this experiment does not display statistical regularity?

2. To investigate the frequencies with which two authors used the equivalent words 'a' and 'an', the total number of occurrences of the two words are recorded in ten excerpts of 100 words each chosen from each of the author's works. The results were:

Sir Arthur Conan Doyle, *Sherlock Holmes stories*: 4, 6, 1, 3, 4, 3, 3, 3, 5, 4
J. R. R. Tolkien, *The Lord of the Rings*: 2, 1, 0, 6, 1, 3, 2, 2, 4, 0

Plot the cumulative relative frequencies of the event 'a randomly chosen word is the word a or an' for both authors separately on the same graph. To what approximate values do these relative frequencies tend?

4 • Equally Likely Outcomes

In some experiments, it is intuitively obvious that all outcomes are equally likely to occur. This chapter describes how to work out probabilities associated with experiments like these.

4.1 Basic ideas

Example 1

The ten digits, $0, 1, \ldots, 9$, are each written on one of ten cards and the cards are thoroughly mixed together in a bag. One card is drawn out, by someone who has been blindfolded, and the digit on it is recorded. A suitable sample space for this experiment is $S = \{0, 1, \ldots, 9\}$.

Suppose we want to find the probability of the event $E =$ 'the outcome is divisible by 3'. We could carry out the experiment a large number of times and determine the probability of E from its relative frequency. But this procedure hardly seems necessary. The symmetry of the experiment suggests that the ten outcomes are all **equally likely** to occur.

In other words, if we did carry out the experiment a large number of times, then we would expect to find that $rf(0) = rf(1) = \ldots = rf(9) = 1/10$. Ideally, then, $P(0) = P(1) = P(2) = \ldots = P(9) = 1/10$. We can express this in another way, by saying that the card is chosen from the bag **at random**. Since $E = \{3, 6, 9\}$ consists of three of the ten different outcomes of the experiment, we would argue that $P(E) = 3/10$.

This is an example of a very important class of experiments. A **finite sample space** is one that consists of finitely many outcomes. Suppose that, in a finite sample space, all k outcomes are equally likely, so each has probability $1/k$. The probability of any event, E, is defined to be

$$P(E) = \frac{\text{number of outcomes in } E}{\text{number of outcomes in } S} = \frac{\text{number of outcomes in } E}{k}$$

In the above experiment, $k = 10$ and $P(E) = 3/10 = 0.3$.

Example 2

(1) Toss a coin, and record the face that lands uppermost. A suitable sample space is $S = \{H, T\}$. The coin is described as **fair** (or **unbiased**) if these two outcomes are equally likely; in other words, if $P(H) = P(T) = 0.5$.

(2) Toss two fair coins, and record the faces that land uppermost. The four outcomes in the sample space $S = \{\mathrm{HH}, \mathrm{HT}, \mathrm{TH}, \mathrm{TT}\}$ are all equally likely. But, suppose that we adopt the sample space $S_1 = \{0, 1, 2\}$ to record simply the

number of tails. The outcomes in this alternative sample space are *not* all equally likely; they have probabilities 0.25, 0.50 and 0.25, respectively.

(3) Roll a standard cubical die, whose faces are marked $1, 2, \ldots, 6$ respectively. If the die is *fair*, then all six possible outcomes have probability $\frac{1}{6}$.

(4) In Scotland in the middle of 1991, there were 299,035 men of pensionable age (i.e. at least 65 years old). Of these, 15,743 were aged 85 years and over. Suppose that we chose a Scotsman of pensionable age *at random* from the population, that is in such a way that everyone was equally likely to be chosen. Then the probability that the man chosen was 85 years or over would have been $15{,}743/299{,}035 = 0.0526$.

To evaluate the probability of an event E in an equally likely outcomes model, we do not need to know which particular outcomes are in E, but just how many of them there are. We could do this by listing them, but this is tedious. This chapter introduces some quicker ways to do the counting.

Example 3

A local restaurant offers a Table d'Hote lunch with three choices of starter, five choices of main course and two choices of dessert. How many different three-course lunches can I have? The answer is $3 \times 5 \times 2 = 30$. This is an example of the use of the Multiplication Principle.

The Multiplication Principle (MP)

Suppose that a complex experiment can be split up into a sequence of steps $\eta_1, \eta_2, \ldots$, and that:

- η_1 has n_1 possible outcomes;
- any outcome of η_1 can be followed by n_2 outcomes of η_2;
- any combination of outcomes of η_1 and η_2 can be followed by n_3 outcomes of η_3;
- and so on.

Then, the compound experiment **η_1 followed by η_2 followed by $\eta_3 \ldots$** has $n_1 n_2 n_3 \ldots$ possible outcomes.

Example 4

A senior quality control inspector has to visit five factories (A, B, C, D and E) every week. He visits a different factory each day, from Monday to Friday, but he chooses the order of his visits *at random* each week. (a) In how many different ways can he order his week's visits? (b) Find the probability that he visits the factories in the same order two weeks running.

SOLUTION

(a) It is worth imagining how the inspector might choose his week's itinerary. He could write the names of the factories on five pieces of paper, mix the papers in a bag, and then draw them out one at a time, visiting the first one on Monday, etc. Define:

η_1: he draws a factory to visit on Monday (five choices);
η_2: he draws one of those remaining to visit on Tuesday (four choices);

η_3: he draws one of those remaining to visit on Wednesday (three choices);
η_4: he draws one of those remaining to visit on Thursday (two choices);
η_5: he draws the remaining factory to visit on Friday (one choice).

By the MP, there are $5 \times 4 \times 3 \times 2 \times 1 = 120$ different ways to order the visits.

(b) By the MP, there are 120×120 different possible ways to choose the order of his visits in two successive weeks. The outcomes are all equally likely. Let G be the event 'he visits the factories in the same order both weeks'. The sequence of procedures that gives an outcome favourable to G is as follows:

η_1': he draws the order for the first week's visits (120 choices);
η_2': he draws the same order for the second week's visits (one choice).

By MP, there are $120 \times 1 = 120$ outcomes in the event G, and so

$$P(G) = \frac{\text{number of outcomes in } G}{\text{number of outcomes in } S} = \frac{120 \times 1}{120 \times 120} = \frac{1}{120}$$

Example 5

In the morning, I have a number of different possible ways to travel from the centre of the city to my office. I could walk all the way, travel by the Underground, or take any of three different buses. Altogether, then, I have $1 + 1 + 3 = 5$ different ways to travel to my office. This is an example of the Addition Principle.

The Addition Principle (AP)

Suppose that the experiments $\eta_1, \eta_2, \ldots$ have (respectively) $n_1, n_2, \ldots$ different possible outcomes. Suppose further that no two of these experiments may be performed simultaneously. The compound experiment $\boldsymbol{\eta_1}$ **or** $\boldsymbol{\eta_2}$ **or ...** has $\boldsymbol{n_1 + n_2 + \ldots}$ possible outcomes.

Example 6

In a class of 25 students (with no twins, triplets, etc.) what is the probability that at least two of the students share the same birthday? To answer this question approximately, it is necessary to make two important simplifying assumptions. The first is to neglect the people who were born on 29 February in a leap year. The second is to assume that equal proportions of people are born on all of the other 365 days of the year.

Each student, then, has one of 365 different possible birthdays. The total number of different (ordered) outcomes for the 25 students is, by the MP, $365 \times 365 \times \ldots \times 365 = 365^{25}$, and (according to the second assumption) these are all equally likely.

Let E = 'at least two students share the same birthday'. We will start by finding out how many outcomes are in the complementary event E' = 'all the students have different birthdays'. There are 365 possible birthdays for the first student, 364 different possible birthdays for the second student, 363 different possible birthdays for the third student, and so on. By the MP, then, there are $365 \times 364 \times 363 \times \ldots \times 341$ different outcomes in E'.

Now any outcome must lie either in E or in E'. Since E and E' cannot be performed simultaneously, it follows from the AP that:

$$\text{no. of outcomes in } E + \text{no. of outcomes in } E' = \text{total no. of outcomes;}$$

that is, no. of outcomes in $E = 365^{25} - (365 \times 364 \times \ldots \times 341)$. So

$$P(E) = \frac{365^{25} - (365 \times 364 \times \ldots \times 341)}{365^{25}} = 1 - \frac{(365 \times 364 \times \ldots \times 341)}{365^{25}}$$
$$= 0.569$$

This is a surprising result at first sight. In fact, the probability of getting at least one match on birthdays in a group of k people is greater than $\frac{1}{2}$ as long as $k \geq 23$. For $k > 40$, the probability is greater than 0.9, and for $k = 50$, the probability reaches 0.970.

EXERCISES ON 4.1

1. A board game uses three cubical dice. The faces of the first die are marked 1, 1, 2, 2, 3 and 3; those of the second die are marked 3, 3, 4, 4, 5 and 5; those of the third die are marked 4, 4, 5, 5, 6 and 6. Write out the 27 different ordered outcomes of the experiment of rolling these three dice. The overall score is found by adding the scores on the second and third dice, and then subtracting the score on the first die. Beside each outcome on your list, write the corresponding overall score. What is the probability of obtaining a score that is an even number?
2. Using the MP, show that a finite set with k elements has exactly 2^k different subsets.
3. A computer has been programmed to generate **random digits**. This means that any digit the computer produces is equally likely to be each of the values $0, 1, \ldots, 9$. Suppose that the computer produces two random digits. Find the probability that they are equal. Suppose that the computer produces three random digits. Find the probability that: (a) they are all equal; (b) they are all different; (c) exactly two of them are equal.

4.2 Permutations

Example 7

Ten athletes are to compete in a race. The published result of the race will consist only of the names of the first three athletes, in the order in which they finish. How many different possible published results are there?

SOLUTION

There are ten possible winners. For each winner, nine athletes might be in second place. For each combination of athletes in first and second place, eight athletes might be in third place. By the MP, then, there are $10 \times 9 \times 8 = 720$ different possible outcomes.

In this example, we were required to choose r out of n distinct objects (athletes), where the *order* in which we chose them was important. The result of such a procedure is called a **permutation** of r from n objects. In general, it is useful to know how many different permutations there are; this number is sometimes written ${}^nP_r$. Applying the MP, as in Example 7, let:

η_1: choose the first object (n ways);
η_2: choose the second object from those remaining ($n-1$ ways);
$\vdots$
η_r: chose the rth object from those remaining ($n-r+1$ ways)

So, the total number of different possible permutations of r from n distinct objects ($r = 1, 2, \ldots, n$) is

$$n \cdot (n-1) \cdot \ldots \cdot (n-r+1)$$

By convention, there is one permutation of 0 from n distinct objects (the case $r = 0$).

In particular, there are $n \cdot (n-1) \cdot \ldots \cdot 1$ distinct permutations of n from n distinct objects (where $n = 1, 2, 3, \ldots$). This number arises so often in practice that we have a special name for it, ***n* factorial**, written $n! = n \cdot (n-1) \cdot \ldots \cdot 1$. For example, $3! = 3 \cdot 2 \cdot 1 = 6$ and $5! = 5 \cdot 4 \cdot 3 \cdot 2 \cdot 1 = 120$. By convention, $0! = 1$. So,

$${}^nP_r = \frac{n!}{(n-r)!} \qquad (r = 0, 1, \ldots, n)$$

In Example 7, $n = 10$ and $r = 3$, so the number of different permutations is

$$\frac{10!}{(10-3)!} = \frac{10 \cdot 9 \cdot 8 \cdot 7 \cdot 6 \cdot 5 \cdot 4 \cdot 3 \cdot 2 \cdot 1}{7 \cdot 6 \cdot 5 \cdot 4 \cdot 3 \cdot 2 \cdot 1} = 10 \cdot 9 \cdot 8 = 720$$

Example 8

An international race is being contested by four Americans, three Britons and three Canadians. If we choose to record the nationality (rather than the name) of the athlete in each position, then there are no longer 10! distinct outcomes. For example, as long as the Americans finish in the first, second, third and tenth position, then we record the same outcome whichever Americans finish in these four positions.

In this example, we wish to find out how many different (or distinguishable) permutations there are of n objects, of which n_1 are alike of one kind, n_2 are alike of another kind, $\ldots$, n_t are alike of a tth kind ($n_1 + n_2 + \ldots + n_t = n$). The answer is

$$\frac{n!}{n_1! n_2! \ldots n_t!}$$

(When $t = n$ and $n_1 = n_2 = \ldots = n_n = 1$, i.e. all the objects are distinct, then this formula shows that there are $n!$ distinct permutations of them, which agrees with the previous result.)

Thus, $n = 10$ (athletes), $t = 3$ (nationalities), $n_1 = 4$ (Americans), $n_2 = 3$ (Britons) and $n_3 = 3$ (Canadians). So, the number of different possible team outcomes is

$$\frac{10!}{4!3!3!} = 4200$$

EXERCISES ON 4.2

1. Calculate the following numbers, without using a calculator: (a) 4!; (b) 7!; (c) 7!/4!; (d) 6!/5!; (e) 3!/0!.
2. How many different permutations are there of:
   (a) five from five distinct objects;
   (b) none from six distinct objects;
   (c) two from eight distinct objects?
3. How many distinct ways are there in which to arrange seven coloured balls in a row, when:
   (a) all seven balls have different colours;
   (b) three of the balls are red, and the rest are all different colours;
   (c) three of the balls are red, two are black, and the rest are all different colours?
4. One suspect and seven other members of the public are taking part in a Police identification parade. In how many different ways may the eight people be arranged in a straight line? If they are assigned an order at random, find the probability that the suspect is either first or last in line.
5. A box of chocolates is to contain two of each of four different chocolates. How many distinct ways are there in which to arrange the chocolates in the eight compartments of the box?

4.3 Combinations

Example 9

The teacher of a class of 11 girls and 9 boys has to choose three of them to represent the class at an inter-school event. (a) In how many different ways can she do this? (b) If she selects three children at random, what is the probability that the group includes at least one girl and at least one boy?

This is a similar problem to Example 7, but there is one big difference. The order in which the teacher chooses the children is unimportant. The group is the same whether she chooses Alan then Beatrice then Claire, or Beatrice then Claire then Alan, or indeed any other permutation of the same three children.

In this kind of example, we wish to choose r out of n objects ($r = 0, 1, \dots, n$), but the order in which they are chosen is not important. An unordered subset of r from n objects is called a **combination**. The total number of different combinations of r from n objects is

$$\frac{n!}{r!(n-r)!}$$

which is denoted $\binom{n}{r}$ and read ***n* choose *r***.

To prove this result, consider any combination. The r elements in the combination can be permuted amongst themselves in $r!$ different ways without changing the combination. So, this one combination corresponds to $r!$ different permutations. It follows that

$$r! \times \binom{n}{r} = {}^nP_r$$

that is,

$$\binom{n}{r} = \frac{n!}{r!(n-r)!} \qquad \text{(see page 25)}$$

Example 9 (continued)

(a) In this case, $n = 20$ and $r = 3$. The total number of different ways of choosing three children, then, is

$$\frac{n!}{r!(n-r)!} = \frac{20!}{3!17!} = \frac{20 \cdot 19 \cdot 18}{1 \cdot 2 \cdot 3} = 1140$$

Notice particularly that the 17! on the denominator cancels all but three of the terms of 20! on the numerator.

(b) If the children are picked at random, then all 1140 different combinations must be equally likely. We now need to find out how many of them correspond to the event G = 'at least one girl and at least one boy are picked'. This event occurs if either one girl and two boys are picked or two girls and one boy are picked.

$$\text{No. of different ways to choose one girl from 11 girls} = \binom{11}{1} = 11;$$

$$\text{No. of different ways to choose two boys from 9 boys} = \binom{9}{2} = \frac{9 \times 8}{1 \times 2} = 36.$$

By the MP, then, the number of different ways to choose one girl and two boys is $11 \times 36 = 396$.

Similarly, the number of different ways to choose two girls and one boy is $\binom{11}{2}\binom{9}{1} = 495$.

By the AP, then, the total number of outcomes in G is $396 + 495 = 891$. So, $P(G) = 891/1140 = 0.782$.

Here are three important results about combinations that follow immediately from the definition.

(1) $\binom{n}{r} = \binom{n}{n-r}$. When r objects are chosen to form a combination, then the $n - r$ objects that are left behind form a combination by default. So, the number of combinations of r objects must equal the number of combinations of $n - r$ objects.

(2) $\binom{n}{0} = \binom{n}{n} = 1$. There is just one combination of n from n objects (since order is unimportant). By convention, there is one way to choose 0 objects from n objects.

(3) $\binom{n}{r} = 0$ when $r < 0$ or $r > n$. There is no way to choose more than n or fewer than 0 objects out of a total of n objects.

$\binom{n}{r}$ is often called a **binomial coefficient**. This is because numbers of this kind appear in the expansion of $(x + y)^n$. For example,

$$(x+y)^2 = x^2 + 2xy + y^2 = \binom{2}{0}x^2y^0 + \binom{2}{1}x^1y^1 + \binom{2}{2}x^0y^2$$

The **Binomial Theorem** states that, for any real numbers x and y, and for any non-negative integer, n,

$$(x+y)^n = \sum_{r=0}^{n} \binom{n}{r} x^{n-r}y^r.$$

EXERCISES ON 4.3

1. How many different combinations are there of r objects from n when: (a) $n = 7, r = 2$; (b) $n = 7, r = 5$; (c) $n = 7, r = 3$; (d) $n = 8, r = 8$?
2. Calculate $\binom{6}{r}$ for $r = 0, 1, \ldots, 6$.
3. Ten athletes are to contest an event in an international competition. Three of them are British. Five of the athletes have to be selected at random to run in heat A of the first round; the remaining five athletes will run in heat B. Find the probabilities that:
   (a) all the Britons run in heat A;
   (b) all the Britons run in the same heat;
   (c) at least one Briton runs in each heat.
4. A quality assessment (QA) team is about to visit a small university department, to assess the quality of its teaching. They will meet four students chosen at random from the honours class of ten students.
   (a) How many different groups of students could be selected to meet the team?
   (b) The class consists of five men and five women. Find the probability that the QA team meet two men and two women.
   (c) Two particular honours students might make a very bad impression: I. Lait, who never attends classes; and Ms R. Abel, who is always complaining. Find the probability that neither of these students is selected to meet the QA team.
5. A domino is a block of wood, card or other material, on whose upper face two numbers appear. For example, here are two typical dominoes.

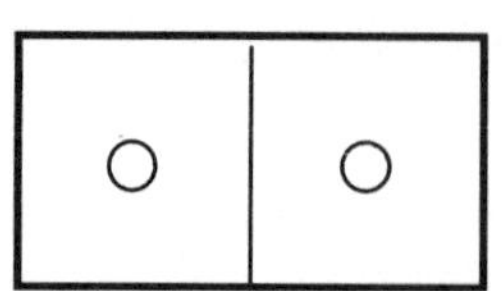

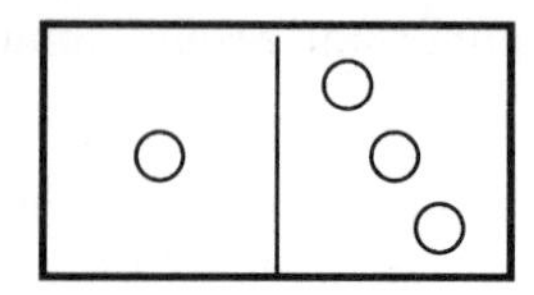

   In a standard set of dominoes, there is every possible pairing of two numbers between zero and six. Each pair appears just once, but (as illustrated above) dominoes with two equal numbers are allowed. Show that there must be 28 dominoes in a standard set.

6. Use the Binomial Theorem to prove that

(a) $\sum_{r=0}^{n} \binom{n}{r} = 2^n;$

(b) $\sum_{r=0}^{n} \binom{n}{r} \theta^r (1-\theta)^{n-r} = 1$, where $0 < \theta < 1$

Application: molecular statistics

Quantum mechanics has revolutionized physics and chemistry by introducing probability models to describe how matter is structured. Electrons, protons and neutrons in an atom, and molecules in a gas, exist in different **quantum states** (energy levels). A fundamental principle of quantum mechanics is that particles of the same kind, or molecules of the same chemical composition, are **indistinguishable**.

To see the practical implications of this, suppose first that we could isolate an assembly consisting of just two identical molecules, A and B. Suppose that A was in quantum state q_1 and that B was in quantum state q_2. The principle of indistinguishability tells us that the observable properties of the assembly would be the same even if the molecules exchanged so that A now had quantum state q_2 and B had quantum state q_1. In a similar way, all $3! = 6$ permutations of three identical molecules among three quantum states must have the same observable properties.

In normal conditions, the number of available quantum states, g, greatly exceeds the number of particles in an assembly, N. The general problem of assigning N molecules to g quantum states is analogous to throwing N indistinguishable balls into g boxes, where $g \gg N$. The MP tells us that, if every box were big enough to contain all N balls, then the number of different ways of throwing N balls into g boxes would be g^N.

For a time it was assumed that, in a gas, all g^N different possible arrangements of N molecules in g quantum states must be equally likely, but they are not. Instead, two quite different models have been found to describe different kinds of particles. These are not competing models; they each fit different situations.

The first model, called **Fermi–Dirac statistics**, describes the arrangement of electrons, protons and neutrons in an atom as well as the molecules in many gases. (As a result, such particles are often called **fermions**.) The model is based on the assumption that no more than one particle is allowed in any given quantum state, a generalization of the **Pauli Exclusion Principle** for electrons.

In our analogy, at most one of the N balls is allowed in each of the g boxes, so exactly N of the boxes must contain one ball each while the remaining $g - N$ boxes must be empty. There are $\binom{g}{N}$ *distinguishable* ways to choose N boxes into which to throw the N indistinguishable balls, and the Fermi–Dirac model states that these combinations are all **equally likely**.

The second model, **Bose–Einstein statistics**, describes the behaviour of assemblies of photons as well as molecules in many gases. (Such particles are consequently called **bosons**.) Any boson can have any of the g available quantum states. Two arrangements of N bosons are distinguishable only if there are different numbers of bosons in at least some quantum states.

In our analogy, each of the g boxes is considered large enough to contain at least N balls, and all the *distinguishable* arrangements of the balls in the boxes are believed to be **equally likely**. Table 4.1 shows all the distinguishable arrangements of $N = 2$ balls (bosons) in $g = 3$ boxes (quantum states). If we label the two balls as A and B, then the MP tells us that there are $3^2 = 9$ different ways to throw the balls into the boxes. Arrangement 1 in Table 4.1 corresponds to just one of these outcomes (A and B both in box 1). Arrangement 2, on the other hand, corresponds to two outcomes (A in box 1, B in box 2; B in box 1, A in box 2). And so on. Despite the fact that the distinguishable arrangements correspond to differing numbers of possible outcomes, the Bose–Einstein model claims that it is the distinguishable arrangements that are all equally likely. Here, then, each distinguishable arrangement has probability $\frac{1}{6}$.

Table 4.1 Distinguishable arrangements of two indistinguishable balls in three boxes

Arrangement	Box 1	Box 2	Box 3
1	**		
2	*	*	
3	*		*
4		**	
5		*	*
6			**

To find out how many distinguishable arrangements there are in general, we pretend that the individual boxes are made up by inserting $g - 1$ interior partitions in a larger box that contains all N balls in a row, for example

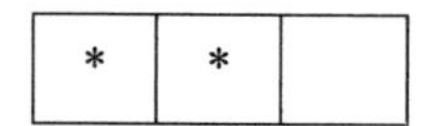

represents the second arrangement in Table 4.1, with two balls and two interior partitions. With $g - 1$ interior partitions and N balls, there are $N + g - 1$ objects to be arranged in a row. There are $\binom{N+g-1}{N}$ different ways to choose positions for the N balls, each corresponding to a distinguishable arrangement of the balls in the boxes. The Bose–Einstein model states that these distinguishable arrangements are all **equally likely**.

For further details, see most university or college-level textbooks on physical chemistry, statistical thermodynamics or statistical mechanics, for example Woodward (1975).

Summary

This chapter has discussed probabilities for finite sample spaces with equally likely outcomes. In such circumstances, the probability of any event is found by dividing the number of outcomes in that event by the total number of equally likely outcomes in the sample space. Various methods have been introduced to make the necessary counting quicker. These include the Multiplication Principle, the Addition Principle, and various results about permutations and combinations. Finally, it has been shown briefly that these concepts underpin modern quantum theories of how subatomic particles and gas molecules are assigned to quantum states.

TUTORIAL PROJECT 4.1

Try out some of the following ways to generate random digits. In each case, check that you obtain roughly the same number of each of the ten digits in the data you generate. (a) Ask 20 different people each to choose a digit 'at random'. (b) Record the last, or second last, digit in 100 telephone numbers found from a telephone directory. (c) Try to find a way to generate random digits using two dice. (Adding together the scores on the dice does not work.) (d) Use a well-shuffled standard pack of playing cards, removing the face cards and counting the aces as ones and the tens as zeros. (e) Find out about the so-called congruential generators used in electronic calculators and computers. The table of random digits given in Appendix 1 was generated in this way; check from a small sample that it does seems to give random digits.

FURTHER EXERCISES

1. Here are some examples of permutations and combinations as they affect modern security systems.
(a) A card allowing you to draw cash from your bank account at an automatic cash dispenser comes with a personal identification number (PIN). The four digits of the PIN must be entered in the correct order before cash can be withdrawn. How many different PIN numbers are there? A thief has three chances to guess a PIN before the machine retains the card; find the probability that the thief guesses the PIN correctly.
(b) The door of a computing room is fitted with a pad, on which are 13 buttons marked $0, 1, \ldots 9$, X, Y and Z. The access code consists of five different characters chosen from the 13 available. The correct five characters may be entered in any order to open the door. How many different possible codes are there?
(c) Access to a computing system is controlled by a password-protection system. A password consists of eight alphanumeric characters (the 26 capital letters of the alphabet and the ten digits), and must begin with a letter. Characters may be repeated as often as required. The characters of the password must be entered in the correct sequence. How many different possible passwords are there? How many passwords would there be if a password could be of any length, from one to eight characters?
2. In the sixteenth and seventeenth centuries, the new subject of probability received an early boost by being used to calculate probabilities associated with popular gambling games. Here are three problems from that era.

(a) Compare the probability of a total score of 9 with that of a total score of 10 when three fair dice are rolled once (Galileo, early seventeenth century).
(b) Compare the probability of obtaining at least one 6 in four rolls of a fair die with that of obtaining at least one double 6 in 24 rolls of two fair dice (Chevalier de Mere, about 1650).
(c) Compare the probability of obtaining at least one 6 when six fair dice are rolled with that of obtaining at least two 6s when 12 dice are rolled (Pepys to Newton, 1693).
[Hint: in (b) and (c), calculate the probabilities of the complementary events.]

3. (From Feller, 1968) Each of n sticks is broken into one long and one short part. The $2n$ parts are randomly regrouped into n pairs, from each of which a new stick is formed. Find the probability that:
(a) all the parts are joined as they were originally;
(b) all the long parts are paired with short parts.
When cells are exposed to harmful radiation, some chromosomes break and behave like these sticks. The 'long' side of the chromosome is the part that contains the centromere. If any two 'long' or any two 'short' sides unite, the cell dies. What is the probability that a cell containing ten chromosomes dies when exposed to radiation?

4. (a) Use the Binomial Theorem to express $(1+x)^{m+n}$ as a power series, where m and n are non-negative integers.
(b) Now write $(1+x)^m$ and $(1+x)^n$ as power series, and so find another power series for $(1+x)^{m+n}$.
(c) Equate the terms in x^s in the two series to show that

$$\binom{m+n}{s} = \sum_{r=0}^{s} \binom{m}{r}\binom{n}{s-r}, \qquad s = 0, 1, \ldots, m+n$$

(d) Interpret this result by considering how to choose a sample of s people from a group of m men and n women.

5. Draw up a table, like Table 4.1, to show all the distinguishable arrangements of three indistinguishable balls in three boxes. What is the probability of the outcome

Box 1	Box 2	Box 3
*	*	*

assuming: (a) a Fermi–Dirac model; (b) a Bose–Einstein model?

5 • The Axioms of Probability

Few experiments have equally likely outcomes. This chapter begins to show how to manipulate probabilities relating to general experiments. A probability model is guaranteed to give consistent results from probability calculations as long as it adheres to four simple axioms. These axioms must hold for any consistent probability model, whether it is derived from a frequentist or a subjectivist point of view. From the Axioms of Probability, a number of further useful results are derived.

5.1 The axioms

The equally likely outcomes models we met in Chapter 4 are examples of **probability models**. In mathematical terms, a probability model is a function, P, which associates with every event, E, a value $P(E)$ in the range $0 \leq P(E) \leq 1$. We call this value the probability of E. Although frequentists and subjectivists disagree about what $P(E)$ means, they do agree about how to make logical deductions within a probability model, from whichever point of view it has been derived.

It is a remarkable fact that a probability model need only obey four simple axioms to be guaranteed to give consistent results from probability calculations. Here are the **Axioms of Probability**, as stated by Kolmogorov (1933).

(1) For any event $E \subseteq S, 0 \leq P(E) \leq 1$.
(2) $P(S) = 1$.
(3) If E and F are *disjoint* events (i.e. $E \cap F = \emptyset$), then $P(E \cup F) = P(E) + P(F)$.
(4) If $E_1, E_2, \ldots$ are *disjoint* events, then $P(E_1 \cup E_2 \cup \ldots) = P(E_1) + P(E_2) + \ldots$.

It might seem as though Axiom 4 is unnecessary given Axiom 3. However, Axiom 3 can be extended by mathematical induction only to cover *finite* sequences of events; Axiom 4 deals with the case of an *infinite* sequence of events.

These axioms reflect similar properties of relative frequencies. If event E occurs on n_E replicates out of a total of n replicates of an experiment, then $rf_n(E) = n_E/n$, which must lie between 0 and 1 (Axiom 1). Also, $rf_n(S) = n/n = 1$ (Axiom 2). If E_1 and E_2 are disjoint, and each E_i occurs on exactly n_i replicates of the experiment, then $E_1 \cup E_2$ occurs on exactly $n_1 + n_2$ replicates, with the result that (Axiom 3)

$$rf_n(E_1 \cup E_2) = \frac{n_1 + n_2}{n} = \frac{n_1}{n} + \frac{n_2}{n} = rf_n(E_1) + rf_n(E_2)$$

Although (most) subjectivists would accept the Axioms of Probability as stated here, they would prefer to work with a modified (conditional) version of them that we shall discuss in the next chapter.

EXERCISE ON 5.1

1. Show that an *equally likely outcomes model* satisfies Axioms 1 to 3. Why do we not need to bother about infinite sequences of disjoint events (Axiom 4) in a finite sample space?

5.2 Some consequences of the axioms

Within a general probability model, as in the equally likely outcomes models discussed in Chapter 4, we want to be able to calculate unknown probabilities from ones we do know. We could perform such calculations directly from the axioms each time, but is much easier to make use of the standard results listed below.

• *Proposition 1*

$P(E') = 1 - P(E)$.

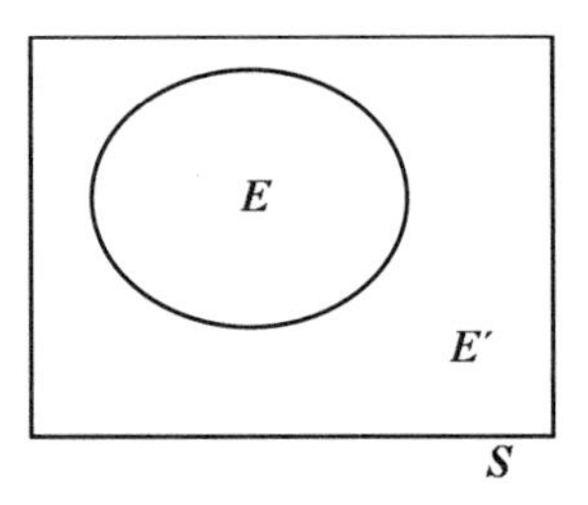

PROOF

$E \cup E' = S$, so

$$P(E \cup E') = P(S) = 1 \quad \text{(Axiom 2)}$$

E and E' are *disjoint*, so

$$P(E \cup E') = P(E) + P(E') \quad \text{(Axiom 3)}$$

Therefore

$$P(E) + P(E') = 1$$

that is,

$$P(E') = 1 - P(E)$$

We have already used Proposition 1 to work out probabilities associated with equally likely outcomes models (see Example 6 in Chapter 4).

Since $\emptyset = S'$, it follows from Axiom 2 and Proposition 1 that $P(\emptyset) = 0$. In other words, **an impossible event has probability 0** (which is intuitively what we would expect). We shall show later (in Chapter 12) that the converse is not true. In other words, $P(E) = 0$ does not necessarily imply that $E = \emptyset$; sometimes, we assign probability 0 to an event that can occur.

• *Proposition 2*

$P(E_1 \cup E_2) = P(E_1) + P(E_2) - P(E_1 \cap E_2)$

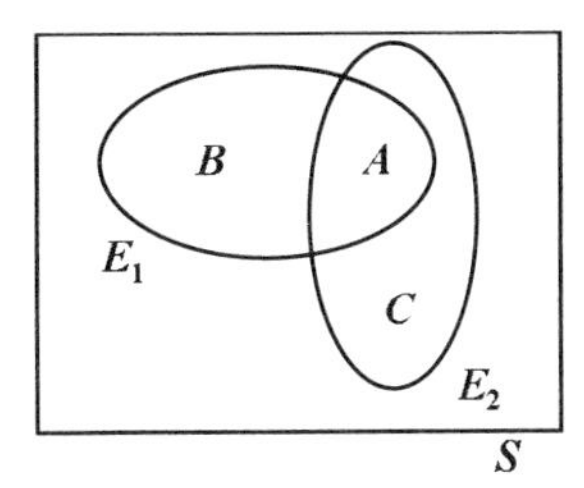

PROOF

$E_1 \cup E_2$ consists of three *disjoint* subsets:

$$A = E_1 \cap E_2$$
$$B = E_1 \cap E_2'$$
$$C = E_1' \cap E_2$$

So, by Axiom 3,

$$P(E_1 \cup E_2) = P(A) + P(B) + P(C)$$

Similarly,

$$P(E_1) = P(A) + P(B) \quad \text{and} \quad P(E_2) = P(A) + P(C)$$

Therefore

$$\begin{aligned} P(E_1) + P(E_2) &= [P(A) + P(B)] + [P(A) + P(C)] \\ &= [P(A) + P(B) + P(C)] + P(A) \\ &= P(E_1 \cup E_2) + P(E_1 \cap E_2) \end{aligned}$$

and so

$$P(E_1 \cup E_2) = P(E_1) + P(E_2) - P(E_1 \cap E_2)$$

Proposition 2 extends Axiom 3, which holds only for *disjoint* events. Notice that, when E_1 and E_2 are disjoint, then $P(E_1 \cap E_2) = P(\emptyset) = 0$, and so this result also gives

$$P(E_1 \cup E_2) = P(E_1) + P(E_2) - P(E_1 \cap E_2) = P(E_1) + P(E_1)$$

Example 1

A house is equipped with two smoke detectors, one downstairs and one upstairs. If a fire starts in the kitchen, which is downstairs, then the downstairs detector will be activated within one minute with probability 0.97, the upstairs detector will be activated within one minute with probability 0.92, and both will be activated within one minute with probability 0.90. Find the probability that, within one minute of a fire starting in the kitchen, at least one of the smoke detectors will be activated within one minute.

SOLUTION

Define the events E_1 = 'downstairs detector is activated within one minute' and E_2 = 'upstairs detector is activated within one minute'. We require to find $P(E_1 \cup E_2)$.

Since $P(E_1) = 0.97$, $P(E_2) = 0.92$ and $P(E_1 \cap E_2) = 0.90$, it follows that

$$P(E_1 \cup E_2) = P(E_1) + P(E_2) - P(E_1 \cap E_2) = 0.97 + 0.92 - 0.90 = 0.99$$

• *Proposition 3*

$$P(E_1 \cup E_2 \cup E_3) = P(E_1) + P(E_2) + P(E_3) - P(E_1 \cap E_2) - P(E_2 \cap E_3) - P(E_3 \cap E_1) + P(E_1 \cap E_2 \cap E_3)$$

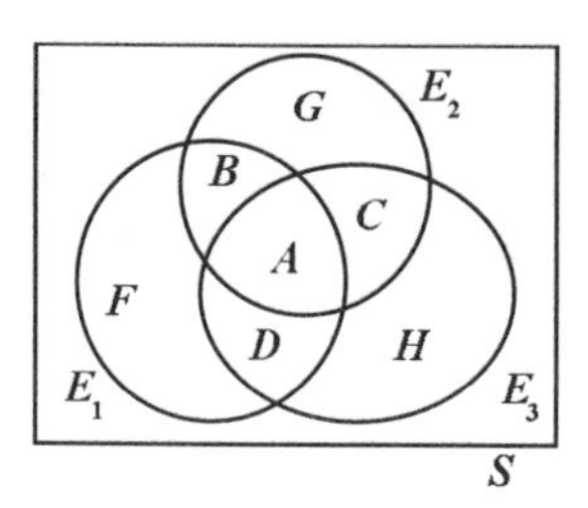

PROOF

The proof is similar to the previous one, using the seven disjoint subsets A, B, C, D, F, G and H shown on the Venn diagram.

Example 2

A survey of shoppers at a chain of cut-price food stores reveals the following purchasing pattern for the three brands of baked beans (X, Y and Z) that the stores stock: 60% of customers sometimes buy brand X, 30% sometimes buy brand Y and 15% sometimes buy brand Z; 10% of customers buy either brands X or Y, depending on availability and price, while another 10% buy either brands X or Z. No customer who would consider buying brand Y would also consider buying brand Z. What proportion of customers do not buy baked beans at these stores?

SOLUTION

Define the events

E_1 = 'a randomly selected shopper sometimes buys X';
E_2 = 'a randomly selected shopper sometimes buys Y';
E_3 = 'a randomly selected shopper sometimes buys Z';
N = 'a randomly selected shopper would not buy any of these brands of baked beans'.

Then $N = (E_1 \cup E_2 \cup E_3)'$.

The survey results tell us all the probabilities required to apply Proposition 3. Since $E_2 \cap E_3$ is the empty set (no customer would buy both brands Y and Z), then it follows that $E_1 \cap E_2 \cap E_3$, which is a subset of $E_2 \cap E_3$, is also the empty set.

Both these events must have probability 0. Hence

$$\begin{aligned} P(E_1 \cup E_2 \cup E_3) &= P(E_1) + P(E_2) + P(E_3) - P(E_1 \cap E_2) - P(E_2 \cap E_3) \\ &\quad - P(E_3 \cap E_1) + P(E_1 \cap E_2 \cap E_3) \\ &= 0.6 + 0.3 + 0.15 - 0.1 - 0.1 - 0 + 0 \\ &= 0.85 \end{aligned}$$

So, $P(N) = 1 - P(N') = 1 - P(E_1 \cup E_2 \cup E_3) = 0.15$. In other words, 15% of all customers never buy any of the three brands of baked beans.

EXERCISES ON 5.2

You might find it helpful to draw an appropriate Venn diagram for some of the following exercises.

1. A survey of the readership of two national newspapers reveals that, in a certain region, 30% of households buy paper A, 35% buy paper B and 45% buy neither newspaper. Show that 10% of households in this region buy both newspapers.
2. Let E_1 and E_2 be any events in a sample space S. Show that the probability that either E_1 or E_2 (but not both) occurs is $P(E_1) + P(E_2) - 2 \cdot P(E_1 \cap E_2)$.
3. Suppose that a finite sample space contains k different outcomes (which are not necessarily equally likely). Show that the probabilities of the k outcomes must sum to 1. [Hint: consider the union of all the elementary events and adapt the method used to prove Proposition 1.]

Summary

This chapter has introduced the Axioms of Probability; as long as a probability model obeys these axioms, then it is guaranteed to assign probabilities consistently to events in the sample space. Arguing from the axioms, some propositions have been proved which allow unknown probabilities to be determined from known ones within a general sample space.

TUTORIAL PROBLEM 1

Look up in the newspapers the betting odds quoted for a horse race or some other event (such as a General Election). Convert the odds for each entrant to the corresponding probability of the entrant winning the event. Use the formula

$$\text{probability} = \frac{\text{odds on}}{1 + \text{odds on}}$$

For example, if the odds are quoted as 2–1 (against), then this is equivalent to 1–2 (on) or 1 : 2 or $\frac{1}{2}$. So, the corresponding probability of winning is $\frac{1}{2}/(1 + \frac{1}{2}) = \frac{1}{3}$. If the odds are 13–2 (against), then the corresponding probability of winning is $\frac{2}{13}/(1 + \frac{2}{13}) = \frac{2}{15}$.

Find the total of all the individual winning probabilities. In the light of Exercise 3 on p. 37, does the result surprise you?

FURTHER EXERCISES

You might find it helpful to draw an appropriate Venn diagram for some of the following exercises.

1. Let E_1 and E_2 be any events in a sample space S. Show that $P(E_1 \cup E_2) \leq P(E_1) + P(E_2)$.
2. (a) Let A and B be events in a sample space S, such that $A \subseteq B$. By writing B as the union of two disjoint events, show that $P(A) \leq P(B)$.
   (b) Now let E_1 and E_2 be any events in S. Show that $P(E_1 \cup E_2) \geq P(E_1)$ and that $P(E_1 \cup E_2) \geq P(E_2)$. [Hint: apply the above result.] How might E_1 and E_2 be related so that $P(E_1 \cup E_2)$ takes its minimum possible value, $P(E_1)$ or $P(E_2)$?
3. (a) Let E_1 and E_2 be events in a sample space S, such that $P(E_1) = 0.3$ and $P(E_2) = 0.5$. Using the results of the last two exercises, find the largest and smallest possible values for $P(E_1 \cup E_2)$.
   (b) Repeat this exercise when $P(E_1) = 0.5$ and $P(E_2) = 0.7$.
4. E_1, E_2 and E_3 are events in a sample space, S, such that $E_1 \cup E_2 \cup E_3 = S$ and $E_1 \cap E_2 \cap E_3 = \emptyset$. If $P(E_1) = 0.25$, $P(E_2) = 0.55$, $P(E_1 \cap E_2) = P(E_1 \cap E_3) = P(E_2 \cap E_3) = 0.1$, find $P(E_3)$.
5. E_1, E_2 and E_3 are events in a sample space, S, such that $P(E_1) = 0.6$, $P(E_2) = 0.2$, $P(E_3) = 0.1$, $P(E_1 \cup E_2) = 0.7$, $P(E_1 \cup E_3) = 0.7$, and $P(E_2 \cup E_3) = 0.3$. Find the probability that none of the events occur.

6 • Conditional Probability and Independence

This chapter introduces two key concepts in probability. First, there is the conditional probability of an event, which takes account of further information that becomes available, for example that some other event is observed to occur. Then, secondly, there is the independence of two events, which means that the fact that one of them happens does not alter the probability that the other happens. Independence, when valid, is a crucial model assumption.

6.1 Conditional probability

Example 1

Table 6.1 contains an extract of data from the Annual Report of the Registrar General for Scotland for the year 1992 (Registrar General, 1993).

Table 6.1 Number of deaths in Scotland in 1992, by age and sex.

	Males		Females		All	
Age range	Population	Deaths	Population	Deaths	Population	Deaths
75–84	91,372	9225	167,714	10,944	259,086	20,169

(*Source*: Registrar General, 1993)

From the last two columns of this table, we can say that a randomly selected person aged between 75 and 84 years, living in Scotland, died in the course of the year 1992 with probability

$$\frac{20{,}169}{259{,}086} = 0.0778$$

This probability does not, however, take account of all the information we have. The table shows that a greater proportion of males than females in this age category died in the course of 1992. The probability that a randomly selected male died was

$$\frac{9225}{91{,}372} = 0.1010$$

which is larger than the overall probability. We call the second value the **conditional** probability that a person aged 75 to 84 died **given** that the person was male.

In general, suppose that E and F are events in a sample space, S, and that $P(F) > 0$. Then the **conditional** probability of E **given** F is defined to be

$$P(E \mid F) = \frac{P(E \cap F)}{P(F)}$$

In Example 1, we looked at the events E = 'a person aged 75–84 years died' and F = 'the person was male'. Since 91,372 of the 259,086 persons in the population were males, and since 9,225 male persons died, then

$$P(F) = \frac{91{,}372}{259{,}086} \quad \text{and} \quad P(E \cap F) = \frac{9225}{259{,}086}$$

By the definition of conditional probability, then

$$P(E \mid F) = \frac{P(E \cap F)}{P(F)} = \frac{9225/259{,}086}{91{,}372/259{,}086} = \frac{9225}{91{,}372}$$

the same result as we obtained before.

The definition of conditional probability is effectively another axiom, in addition to the four Axioms of Probability stated in Chapter 5.

We cannot 'prove' an axiom, but a relative frequency argument can help us to see why this definition of conditional probability produces reasonable answers. Suppose that the events E, F and $E \cap F$ occur on (respectively) r_E, r_F and $r_{E \cap F}$ out of n replicates of an experiment. So, $rf_n(E) = r_E/n$, for example. Now E must occur on exactly $r_{E \cap F}$ of the r_F replicates on which F occurred. In other words, the **conditional relative frequency** of E **given** that F occurs is $r_{E \cap F}/r_F$. In the limit, then, we can define

$$P(E \mid F) = \lim_{n \to \infty} \frac{r_{E \cap F}}{r_F} = \lim_{n \to \infty} \frac{r_{E \cap F}/n}{r_F/n} = \frac{\lim_{n \to \infty} r_{E \cap F}/n}{\lim_{n \to \infty} r_F/n} = \frac{P(E \cap F)}{P(F)}$$

Example 2

A bag contains 50 balls that are identical apart from colour; 35 of them are black and the other 15 red. Consider the experiment of drawing balls from the bag at random and without replacement. (a) Find the probability that the first ball drawn is red. (b) *Given* that the first ball drawn is black, find the *conditional* probability that the second ball is red. (c) Find the probability that the first ball drawn is black *and* the second ball drawn is red.

SOLUTION

(a) Since the ball is drawn at random, there are 50 equally likely outcomes. So,

$$P(\text{first ball drawn is red}) = 15/50 = 0.3$$

(b) Given that a black ball has been drawn and not replaced, 49 balls remain in the bag, of which 34 are black and 15 are red. The second ball is to be drawn at random, so

$$P(\text{second ball drawn is red} \mid \text{first ball drawn is black}) = 15/49 = 0.306$$

(c) Using the Multiplication Principle (Chapter 4)

no. of possible ways to draw two balls (without replacement) = 50×49
no. of favourable outcomes = 35×15

Therefore required probability = $(35 \times 15)/(50 \times 49) = 0.2143$.

EXERCISES ON 6.1

1. The table below extends Table 6.1, to give information about the number of deaths of all persons aged 65 years old and over in Scotland in the year 1992. For each age group, calculate the overall probability of death in the course of the year. Then calculate the conditional probability of death for males and females separately in each age group.

	Males		Females		All	
Age range	Population	Deaths	Population	Deaths	Population	Deaths
65–74	191,920	8541	249,174	6663	441,094	15,204
75–84	91,372	9225	167,714	10,944	259,086	20,169
85+	15,743	3556	52,959	9008	68,702	12,564

(*Source*: Registrar General, 1993)

2. A student moving into a new flat buys a box of 20 light bulbs. Unknown to the student, exactly one of the bulbs is defective. Find:
(a) the probability that the first bulb used is not defective;
(b) the *conditional* probability that the second bulb used is not defective, *given* that the first bulb used is not defective;
(c) the *conditional* probability that the second bulb used is not defective, *given* that the first bulb used is defective.

6.2 The Multiplication Theorem of probability

There is a more general way to solve problems like part (c) of Example 2. From the definition of conditional probability, it follows that, for any events E and F such that $P(F) > 0$,

$$P(E \cap F) = P(E \mid F) . P(F)$$

This general result is sometimes known as the **Multiplication Theorem** of probability. In Example 2(c), letting E = 'second ball is red' and F = 'first ball is black', then

$$P(F) = 35/50$$

since 35 out of the 50 balls are black, and

$$P(E \mid F) = 15/49$$

since 15 out of the 49 balls left are red; and so

$$P(E \cap F) = P(E \mid F).P(F) = 15/49 \times 35/50$$

Example 3

Doctors sometimes refer to the 'rule of halves' when discussing the prevalence of common chronic disorders, such as hypertension (high blood pressure) and non-insulin-dependent diabetes. This 'rule' claims that only $\frac{1}{2}$ of all cases of a chronic condition are detected, only $\frac{1}{2}$ of those detected are treated and only $\frac{1}{2}$ of those treated are controlled (Hart, 1992).

Define the events E_1 = 'disorder is detected', E_2 = 'disorder is treated' and E_3 = 'disorder is controlled'. Then the rule of halves suggests that

$$P(E_1) = \tfrac{1}{2} \quad P(E_2 \mid E_1) = \tfrac{1}{2} \quad P(E_3 \mid E_2 \cap E_1) = \tfrac{1}{2}$$

What is the probability that a case is detected *and* treated *and* controlled? In other words, what is $P(E_3 \cap E_2 \cap E_1)$? Extending the Multiplication Theorem, we can write

$$P(E_3 \cap E_2 \cap E_1) = P(E_3 \mid E_2 \cap E_1) \cdot P(E_2 \mid E_1) \cdot P(E_1) = \tfrac{1}{2} \cdot \tfrac{1}{2} \cdot \tfrac{1}{2} = \tfrac{1}{8}$$

This suggests that only one person in eight who suffers from a chronic disorder has that condition controlled by appropriate long-term treatment. Hart (1992) reviews some research evidence in favour of this conclusion, though the purpose of this paper is to provide a stimulus to increase the proportion greatly.

Example 3 has suggested how to extend the Multiplication Theorem to three events. In fact, it can be extended to m events, as follows. Suppose that $E_1, E_2, \ldots, E_m$ are events such that $P(E_{m-1} \cap \ldots \cap E_2 \cap E_1) > 0$. Then

$$P(E_m \cap \ldots \cap E_1) = P(E_m \mid E_{m-1} \cap \ldots \cap E_1) \times P(E_{m-1} \mid E_{m-2} \cap \ldots \cap E_1) \times \ldots \times P(E_2 \mid E_1) \times P(E_1)$$

You might like to attempt to prove this result for yourself, using a proof by induction on m. Notice that (for $m \geq 3$)

$$E_m \cap \ldots \cap E_1 = E_m \cap (E_{m-1} \cap \ldots \cap E_1)$$

EXERCISES ON 6.2

1. A bag contains 100 balls that are identical apart from colour. 40 balls are black, 40 red and 20 blue. Balls are to be drawn from the bag at random and without replacement. Find the probabilities of the following events:
   (a) the first ball drawn is black;
   (b) the first ball drawn is black and the second ball drawn is red;
   (c) there are no blue balls among the first three drawn.
2. Before sending a consignment of 50 desktop computers to a valued customer, a manufacturer chooses a few computers at random and subjects them to stringent tests. Unknown to the manufacturer, two of the 50 computers have defects. What is the probability that none of the first three computers chosen for test will be defective?

3. Of the patients admitted to general practitioner beds in the community hospitals of a certain English health authority, 12% die while admitted. Of those who survive, 75% return to their own or a relative's home. What percentage of all patients admitted to these beds return to their own or a relative's home?

6.3 Independence

One of the most crucial concepts in probability is that of the independence of two events. Informally, we say that E and F are independent events if the occurrence or non-occurrence of F does not change the probability of E (and vice versa). Formally, the events E and F are defined to be **independent** if

$$P(E \cap F) = P(E) \times P(F)$$

Whenever $P(F) > 0$, so that $P(E \mid F)$ is sensibly defined,

$$\begin{aligned} & P(E \cap F) = P(E) \times P(F) \\ \Leftrightarrow\ & \frac{P(E \cap F)}{P(F)} = P(E) \\ \Leftrightarrow\ & P(E \mid F) \ = P(E) \end{aligned}$$

In other words, when $P(F) > 0$, then E and F are independent if and only if the *conditional* probability of E *given* F is the same as the *unconditional* probability of E. Similarly, if $P(E) > 0$, then E and F are independent if and only if the *conditional* probability of F *given* E is the same as the *unconditional* probability of F. Intuitively, these are precisely the relationships we want the formal definition of independence to express.

In practice, it is very rare to discover that two events are independent by working out separately the probabilities of E, F and $E \cap F$ and showing that they are related by the above formula. Almost invariably, as in the following examples, independence is a *model assumption* justified by our knowledge of experimental conditions. When we can be sure that events are independent, even apparently complex probabilities become relatively easy to calculate. It is crucial to ensure that an independence assumption is realistic; otherwise, the answers produced will often be quite misleading.

Example 4

A gas central-heating system consists of a pump and a boiler in series; it breaks down if either component fails.

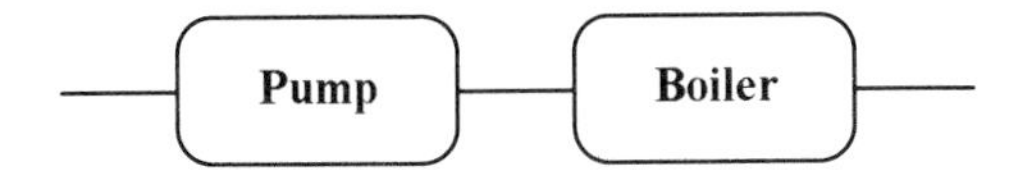

In any year, the pump and boiler (respectively) fail with probabilities θ_1 and θ_2. Failures in the two components are assumed to occur *independently*. Find the probability that the central-heating system breaks down in a given year.

SOLUTION

Let E = 'pump fails' and F = 'boiler fails'. Then

$$\begin{aligned} P(\text{system breaks down}) &= P(E \cup F) \\ &= P(E) + P(F) - P(E \cap F) \qquad \text{(Chapter 5)} \\ &= P(E) + P(F) - P(E) \cdot P(F) \quad \text{(Independence)} \\ &= \theta_1 + \theta_2 - \theta_1 \cdot \theta_2 \end{aligned}$$

We can extend the concept of independence to more than two events. For example, the three events E, F and G are said to be *independent* if

(1) they are pairwise independent (i.e. E and F are independent, E and G are independent, F and G are independent); and
(2) $P(E \cap F \cap G) = P(E) \cdot P(F) \cdot P(G)$.

Notice that both these conditions are necessary, since neither necessarily implies the other (see Exercises 3 and 4 on pp. 45–6).

Example 5

In this example, a symptomatic patient is someone who attends his or her doctor with symptoms suggestive of conjunctivitis (a disorder of the eyes). On the basis of years of experience, a consultant believes that a symptomatic patient has conjunctivitis with probability $1 - 3\theta$. The consultant further believes that symptomatic patients might have one of three other disorders (A, B and C), each with probability θ.

In search of further evidence for the model, the consultant checks through the records of N patients who have recently attended their doctor with symptoms of conjunctivitis. Find the probability that the consultant discovers at least one case of each of the disorders A, B and C.

SOLUTION

Define the events E_1 = 'the consultant finds no case of disorder A', E_2 = 'the consultant finds no case of disorder B', E_3 = 'the consultant finds no case of disorder C' and F = 'the consultant finds at least one case of each of the three disorders'. Since $F = (E_1 \cup E_2 \cup E_3)'$, we require to find $P(F) = 1 - P(E_1 \cup E_2 \cup E_3)$. We shall do this using Proposition 3 on page 36.

The probability that a patient does not have disorder A is $(1 - \theta)$. So, $P(E_1) = P(\text{none of the } N \text{ patients has disorder A}) = (1 - \theta)^N$, assuming that the patients' outcomes were all independent. Similarly, $P(E_2) = P(E_3) = (1 - \theta)^N$.

The probability that a patient has neither disorder A nor disorder B is $(1 - 2\theta)$. So, $P(E_1 \cap E_2) = P(\text{none of the } N \text{ patients had disorder A or disorder B}) = (1 - 2\theta)^N$, assuming that the patients' outcomes were all independent. Similarly, $P(E_1 \cap E_3) = P(E_2 \cap E_3) = (1 - 2\theta)^N$.

Finally, $P(E_1 \cap E_2 \cap E_3) = (1 - 3\theta)^N$, by a similar argument. Therefore,

$$P(F) = 1 - P(E_1 \cup E_2 \cup E_3) = 1 - [3(1 - \theta)^N - 3(1 - 2\theta)^N + (1 - 3\theta)^N]$$

If two *disjoint* events, E and F, have non-zero probabilities, then they cannot be independent. This is intuitively obvious, since knowing that F has occurred tells us that E cannot occur. Formally, $P(E \cap F) = 0$ whereas $P(E), P(F) > 0$, and so $P(E \cap F) \neq P(E) \times P(F)$. Table 6.2 summarizes our results about disjoint and independent events.

Table 6.2 Some probability results for disjoint and independent events.

Events E and F such that $P(E) > 0$ and $P(F) > 0$		
E and F disjoint	E and F not disjoint	
	E and F independent	E and F not independent
$P(E \cap F) = 0$	$P(E \cap F) = P(E) \cdot P(F)$	
$P(E \cup F) = P(E) + P(F)$	$P(E \cup F) = P(E) + P(F) - P(E) \cdot P(F)$	$P(E \cup F) = P(E) + P(F) - P(E \cap F)$
$P(E \mid F) = 0$	$P(E \mid F) = P(E)$	$P(E \mid F) = P(E \cap F)/P(F)$
$P(F \mid E) = 0$	$P(F \mid E) = P(F)$	$P(F \mid E) = P(F \cap E)/P(E)$

EXERCISES ON 6.3

1. A (possibly biased) coin is tossed twice. The result 'heads' is noted if the coin lands heads up on the first toss and tails up on the second toss (HT). In a similar way, the result 'tails' is noted if the outcome is TH. HH and TT are ignored, the entire experiment being repeated until either HT or TH occurs. Show that the result 'heads' is noted with probability 0.5 (however biased the coin).
2. A ship is equipped with two propulsion engines. It normally operates with both engines running, but can proceed almost as normal as long as one of the engines is working. In the course of a typical voyage, each engine (independently) will fail with probability θ ($0 < \theta < 1$). We say that the engines are running in parallel (see the diagram below).

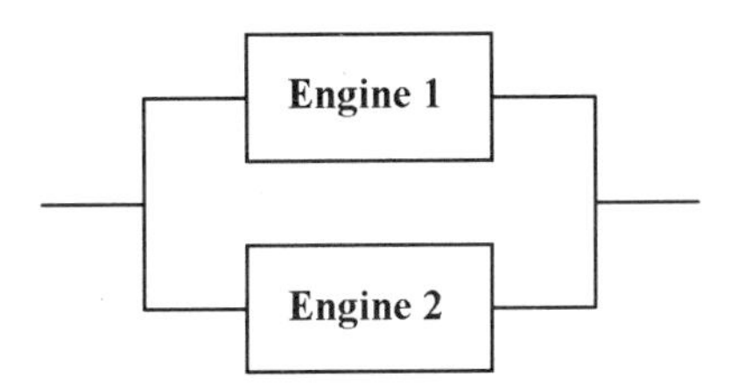

   Show that the probability that both engines fail in the course of a typical voyage is smaller than θ. How realistic, do you think, is the assumption that failures occur independently in the two engines?
3. (Feller, 1968) Two fair dice are tossed. Let A = 'the score on the first die is an odd number', B = 'the score on the second die is an odd number' and C = 'the total score on the two dice is an odd number'. Write out the 36 ordered outcomes and, beside each, write the total score. Assuming equally likely outcomes, show that these three events are pairwise independent but not independent.

4. The probabilities of the scores $1, 2, \ldots, 6$ on a biased die are 1/16, 2/16, 2/16, 3/16, 2/16 and 6/16 respectively. Define the events $A = \{1, 2, 3, 4\}$, $B = \{1, 3, 4, 5\}$, $C = \{3, 6\}$. Show that $P(A \cap B \cap C) = P(A) \cdot P(B) \cdot P(C)$ but that the three events are not independent.

Summary

This chapter has introduced two key concepts: the conditional probability of one event given another event, and the independence of two events. The Multiplication Theorem of probability has also been introduced, and extended for use with more than two events.

FURTHER EXERCISES

1. E_1 and E_2 are events in a sample space, S, such that $P(E_1) = 0.4$ and $P(E_1 \cup E_2) = 0.6$. Find $P(E_2)$ if:
   (a) E_1 and E_2 are disjoint;
   (b) E_1 and E_2 are independent;
   (c) $P(E_2 \mid E_1) = 0.2$;
   (d) $P(E_1 \mid E_2) = 0.5$.
2. Students on a certain university course may fail to complete it for three reasons: 20% of all students who enrol on the course withdraw from it voluntarily; of those who do not withdraw, 5% fail because of unsatisfactory classwork; of those whose classwork is satisfactory, 10% fail the final exam. What overall percentage of students enrolling on this course fail to complete it?
3. A particular circuit on an aircraft fire-warning system only operates as long as both a C1 and a C2 component continue to work. A C1 component fails in the course of a flight with probability θ_1 $(0 < \theta_1 < 1)$, a C2 component fails with probability θ_2 $(0 < \theta_2 < 1)$. Failures are assumed to occur independently in all components.
   (a) The basic circuit is shown below. Find the probability that the circuit fails during a given flight.

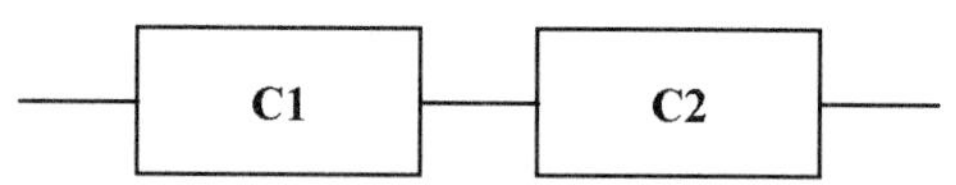

   (b) Because of the circuit's importance to the aircraft's safety, it has been suggested that its components should be duplicated as shown below. Each branch of the circuit fails if either its C1 or C2 component fails. The circuit now fails only if both branches fail. Find the probability that this circuit fails during a given flight, and show that this probability is smaller than that calculated in (a).

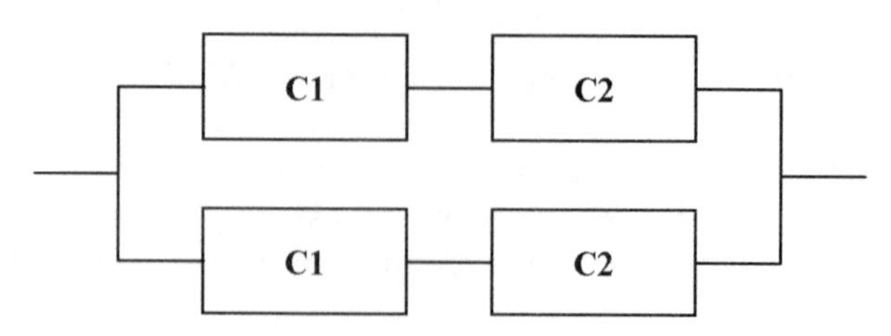

(c) As an alternative, the individual components are duplicated, as shown below. Now the system continues to function until either both C1 components or both C2 components fail. Find the probability that this circuit fails during a given flight, and show that it is smaller than the probability calculated in (b).

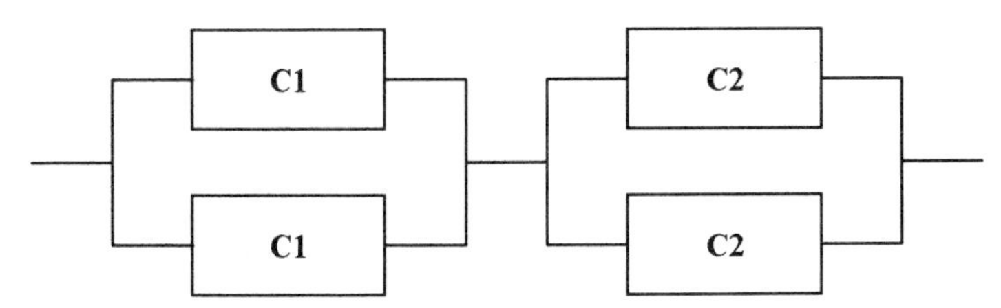

4. In a sale, you buy five new videos. One of these, though you do not know it yet, is of poor technical quality. You choose at random the order in which you intend to play the five videos. Find the probability that the defective video is:
(a) the first that you intend to play;
(b) the second that you intend to play;
(c) the third that you intend to play.
Do you find these answers surprising?
5. Working from the Axioms of Probability and the definition of conditional probability, show that the following results must hold for any fixed event G such that $P(G) > 0$. [You will need to use the distributive laws of set theory, see Section 2.4.]
(a) For any events $E \subseteq S$, $0 \leq P(E \mid G) \leq 1$.
(b) $P(G \mid G) = 1$.
(c) If $E_1, E_2, \ldots$ are events such that $E_1 \cap G, E_2 \cap G, \ldots$ are *disjoint*, then $P(E_1 \cup E_2 \cup \ldots \mid G) = P(E_1 \mid G) + P(E_2 \mid G) + \ldots$.
(d) For any events E and $F \subseteq S$ such that $P(F \cap G) > 0$, $P(E \mid F \cap G)P(F \mid G) = P(E \cap F \mid G)$.
Some probabilists would prefer to use axioms equivalent to these results, instead of the axioms stated in Chapter 5, as a starting point for mathematical probability (see Lindley, 1965).

7 • Bayes' Theorem

Chapter 6 introduced the Multiplication Theorem to work out $P(E \cap F)$ from $P(E \mid F)$ and $P(F)$. This chapter describes the Law of Total Probability which extends that result. It also introduces Bayes' Theorem, which allows the (conditional) probability of an event to be updated in the light of new information.

7.1 The Law of Total Probability

Example 1

Down's syndrome is a genetic disorder that affects about 0.15% of all British babies. The amniocentesis test for Down's syndrome, administered during pregnancy, identifies very accurately whether or not a baby will be born with this disorder. Unfortunately, a small proportion of foetuses die as a result of having the test. So, less accurate, but safe, screening tests are being developed. Only a child whose result on such a test is positive would be sent for amniocentesis. Although tests of this kind are entirely voluntary, it is assumed that almost all mothers will accept the screening test if it carries no risk to the baby.

In a recent study of one such test, it has been found that 60% of babies with Down's syndrome, and 6% of babies who do not have this disorder, test positive. What percentage of all children will be referred for amniocentesis if this screening test is put into routine use during all British pregnancies?

Before answering this question, we need to develop some terminology. Let E_1 be the event 'a randomly selected baby has Down's syndrome' and E_2 the event 'a randomly selected baby does not have Down's syndrome'. If we consider the experiment of examining any child, either E_1 or E_2 (but not both) must be true. We say that these events **partition** the sample space for this experiment.

• Definition

A collection of events $E_1, E_2, \ldots$ forms a **partition** of the sample space S if:

(1) $\bigcup_i E_i = S$;

(2) $E_i \cap E_j = \emptyset$ when $i \neq j$;

(3) $P(E_i) > 0$ for each i.

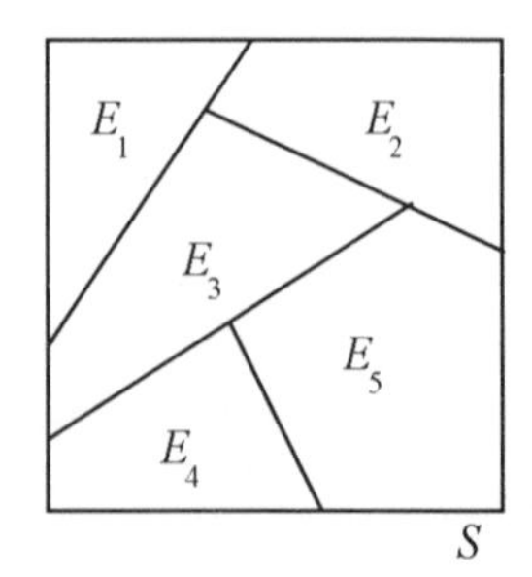

We want to find the probability of the event A = 'a randomly selected baby tests positive'. The study has told us that $P(A \mid E_1) = 0.60$ and that $P(A \mid E_2) = 0.06$. The following result allows us to combine these conditional probabilities with $P(E_1)$ and $P(E_2)$ to find $P(A)$.

The Law of Total Probability

Suppose that the events $E_1, E_2, \ldots$ partition the sample space S. Let A be any event in S. Then

$$P(A) = \sum_i P(A \mid E_i) \cdot P(E_i)$$

PROOF

It follows from the distributive law of set theory (Section 2.4) that

$$A = A \cap S = A \cap \left(\bigcup_i E_i \right) = \bigcup_i (A \cap E_i)$$

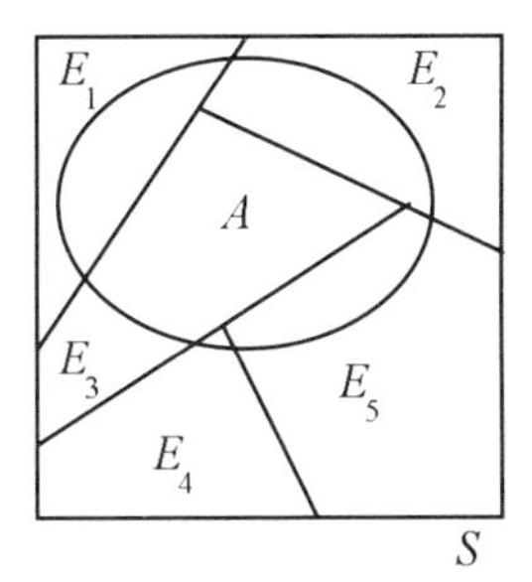

Since the E_i are disjoint, the $(A \cap E_i)$ are disjoint too. So,

$$\begin{aligned} P(A) &= \sum_i P(A \cap E_i) && \text{(by Axiom 4)} \\ &= \sum_i P(A \mid E_i) \cdot P(E_i) && \text{(by definition of conditional probability)} \end{aligned}$$

Applying this result to Example 1,

$$\begin{aligned} P(A) &= P(A \mid E_1) \cdot P(E_1) + P(A \mid E_2) \cdot P(E_2) \\ &= (0.60 \times 0.0015) + (0.06 \times 0.9985) \\ &= 0.0009 + 0.0599 \\ &= 0.0608 \end{aligned}$$

that is, 6.08% of all children test positive and will be referred for amniocentesis.

Drawing a **tree diagram** helps us keep track of the probabilities in examples like this one. Figure 7.1 shows a tree diagram for Example 1, indicating all the possible outcomes and the associated (unconditional and conditional) probabilities. Notice that the unconditional probabilities are on the left-hand branches of this diagram.

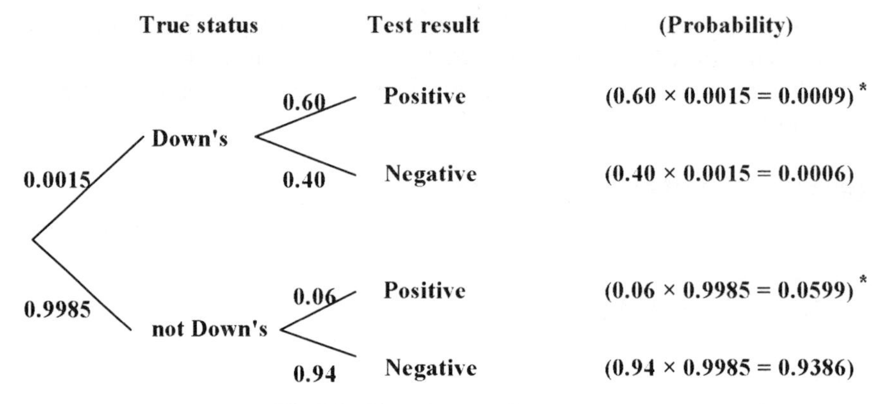

Fig 7.1 Tree diagram for Example 1.

In the probability column on this diagram, the first entry (for example) is the probability that a child has Down's syndrome *and* tests positive. Summing the entries marked * in this column gives P(child tests positive)[$= P(A)$] $= 0.0608$ as before.

Example 2

Imagine planning a sample survey to discover what proportion of British university students have ever used illegal drugs. If asked the question outright, many students might lie to avoid embarrassment. So we prepare three question cards. The first says 'Have you ever used illegal drugs?' The second card instructs the student to 'Answer No' and the third instructs the student to 'Answer Yes'. We intend to show each respondent all three cards, shuffle them, allow the respondent to choose one at random (without showing it to us) and then to respond appropriately. In this way we hope to persuade virtually everyone to respond truthfully. But how do we then find out the proportion of students who have used illegal drugs?

SOLUTION

Let $\theta = P$ (a randomly selected student has used illegal drugs). The tree diagram for this experiment is shown in Fig. 7.2.

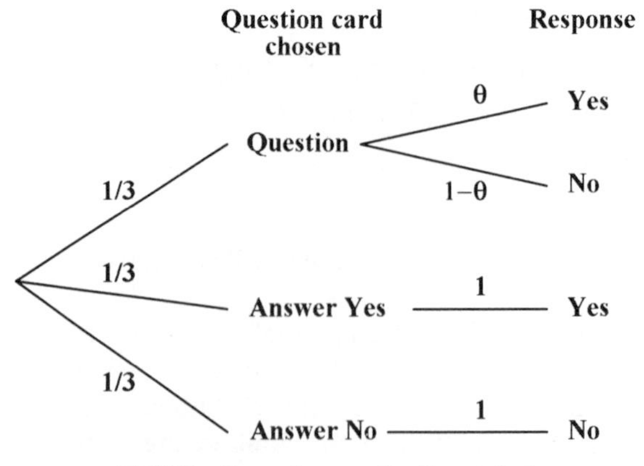

Fig 7.2 Tree diagram for Example 2.

Letting A be the event 'Respondent answers Yes', then

$$P(A) = P(A \mid E_1) \cdot P(E_1) + P(A \mid E_2) \cdot P(E_2) + P(A \mid E_3) \cdot P(E_3)$$
$$= \theta \cdot 1/3 + 1.1/3 + 0.1/3$$
$$= (\theta + 1)/3$$

If, out of a large number of students questioned (n, say), a total of r respond Yes, then

$$P(A) \cong r/n \quad \text{(relative frequency definition of probability)}$$

that is,

$$(\theta + 1)/3 \cong r/n$$

or

$$\theta \cong 3r/n - 1$$

EXERCISES ON 7.1

1. Clefting is a congenital disorder with two distinct forms: (1) isolated cleft palate (CP) and (2) cleft lip, with or without cleft palate (CL(P)). In the UK as a whole, 30% of all cases are CP. About two-thirds of CP cases are females. The position is reversed with CL(P), where about two-thirds of cases are males. Overall, what proportion of UK clefts occur in females?

   In the West of Scotland population, CP and CL(P) are equally common. In the West of Scotland, then, what proportion of all clefts occur in females?
2. Asthma is the commonest cause of a British child being hospitalized. It is a growing problem. In the decade of the 1980s, the annual number of hospital admissions due to asthma trebled. A number of recent studies have demonstrated that, though this is partly due to more children becoming asthmatic, it is also a result of increasing readmission rates.

   The probability that a child will be readmitted to hospital within six months of a given admission depends on how often they have been admitted previously. A recent study found that only 20% of first admissions were readmitted within 6 months, compared with 25% of those who had already been admitted once before, 33% of those who had been admitted twice previously, 40% of those who had been admitted three times previously and 50% of those who had previously been admitted more often than three times.

   In a local hospital, in a typical period, 46%, 25%, 9%, 5% and 15% of the children admitted because of asthma attacks have been admitted 0, 1, 2, 3 and 4 or more times previously (respectively). What overall percentage of these children will be readmitted within six months?

7.2 Bayes' Theorem

Now let us go back to Example 1 and consider a further question. Of all the children whose test is positive, what proportion actually have Down's syndrome? We would think that a very large proportion do, but we would be wrong. In fact the answer is just 15 in every 1000! To prove this, we need a further result known as Bayes' Theorem.

Bayes' Theorem

Suppose that the events $E_1, E_2, \ldots$ partition the sample space S. Let A be any event in S, with $P(A) > 0$. Then

$$P(E_j \mid A) = \frac{P(A \mid E_j) \cdot P(E_j)}{\sum_i P(A \mid E_i) \cdot P(E_i)} \quad (j = 1, 2, \ldots)$$

PROOF

By the Multiplication Theorem of probability (Section 6.2),

$$P(A \cap E_j) = P(A \mid E_j) \cdot P(E_j)$$

and

$$P(A \cap E_j) = P(E_j \cap A) = P(E_j \mid A) \cdot P(A)$$

It follows that $P(E_j \mid A) \cdot P(A) = P(A \mid E_j) \cdot P(E_j)$, that is

$$P(E_j \mid A) = P(A \mid E_j) \cdot P(E_j)/P(A)$$

Baye's Theorem now follows by the Law of Total Probability.

In Example 1, the *conditional* probability that a child has Down's syndrome, *given* that the child is sent for amniocentesis, is

$$\begin{aligned} P(E_1 \mid A) &= (0.60 \times 0.0015)/[(0.60 \times 0.0015) + (0.06 \times 0.9985)] \\ &= 0.0009/[0.0009 + 0.0599] \\ &= 0.0148 \end{aligned}$$

In other words, just 1.48% of all children sent for amniocentesis following the proposed screening test will be found to have Down's syndrome. Another way to look at this is to say that almost 99% of children referred for amniocentesis will be found not to have Down's syndrome after all.

This is still an improvement on a system of carrying out the amniocentesis test during all pregnancies, when just 0.15% of babies tested will have Down's syndrome. We can think of this as meaning that, before the test is carried out, the **prior** probability that any baby has Down's syndrome is 0.0015 (the population proportion). Once a positive test result is obtained, the updated (or **posterior**) probability that the baby has Down's syndrome is 0.0148. Bayes' Theorem, then, provides a way to incorporate new information into the probability assigned to an event.

EXERCISES ON 7.2

1. Referring to Exercise 1 on p. 51, what proportion of females with clefts have isolated CP in (a) the UK as a whole, (b) the West of Scotland?
2. Referring to Exercise 2 on p. 51, given that a person is readmitted within six months, what is the conditional probability that the person was not a first admission?

Application: simple genetics

Human beings each have one of two blood types, Rh positive and Rh negative. Like many other traits, our blood type is determined by just one gene. This gene appears in two forms (or **alleles**), which we shall denote by R and r. One or other allele of this gene appears on each of an individual's two chromosomes. So, there are three **genotypes**:

(1) RR (individuals with an R gene on both chromosomes);
(2) rr (individuals with an r gene on both chromosomes);
(3) Rr (individuals with an R gene on one chromosome and an r gene on the other, the order being unimportant).

We inherit our R or r genes from our parents according to the following probabilistic rules. Parents each donate one gene to a child. Each parent is *equally likely* to donate each gene he or she possesses to a given child (and this process is *independent* for different children). Parents donate genes to a given child *independently* of one another. For example, the tree diagram in Fig 7.3 indicates that the child of an RR and an Rr parent is equally likely to be RR and Rr.

Genotypes RR and Rr have the same appearance or **phenotype** (Rh-positive blood). Only genotype rr has a different appearance (Rh-negative blood). This effect is called **complete dominance**, R being the **dominant**, and r the **recessive**, form of the gene. We speak of an RR individual as a **pure dominant**, rr as a **pure recessive**, and Rr as a **hybrid**.

Figure 7.4 shows that two hybrid parents can produce children of any genotype: RR with probability 0.25, Rr with probability 0.5 and rr with probability 0.25.

This means that at least some pairs of Rh-positive parents (both hybrids) can produce Rh-negative children. Can parents who are both Rh-negative produce Rh-positive children?

Among the many human characteristics passed on genetically are certain well-known *disorders* (e.g. red–green colour blindness, albinism, haemophilia, some forms of muscular dystrophy, sickle-cell anaemia). These are much less common among females than among males. For example, colour blindness affects 5% of males and just 0.25% of females in the UK. Why should this be?

The gene controlling colour blindness (B, b) is said to be **sex-linked** or **X-linked**. It differs from the **autosomal** gene discussed previously in appearing only on X chromosomes. Females have two X chromosomes, and so have three possible genotypes as before (BB, Bb and bb). Males, though, have one X and one Y chromosome, and so have just two possible genotypes, B_ and b_.

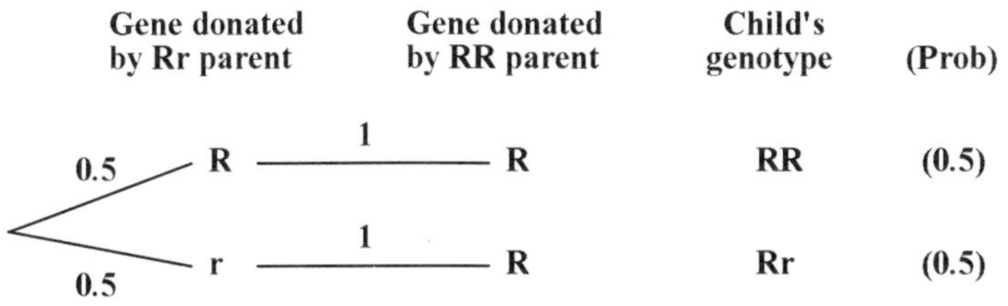

	Gene donated by Rr parent		Gene donated by RR parent	Child's genotype	(Prob)
0.5	R	1	R	RR	(0.5)
0.5	r	1	R	Rr	(0.5)

Fig 7.3 Tree diagram showing the possible genotypes of the offspring of a hybrid (Rr) and a pure dominant (RR) parent.

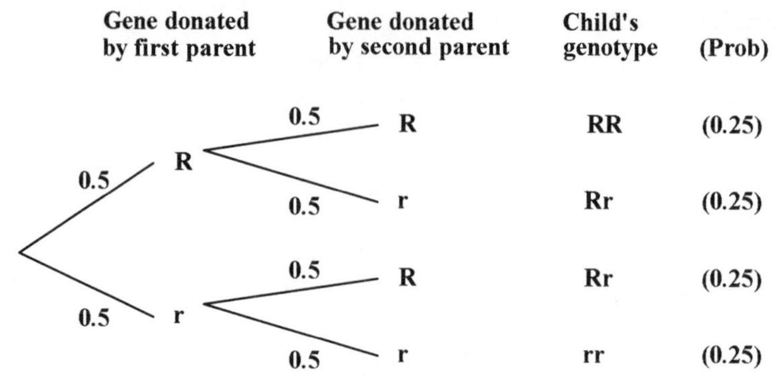

Fig 7.4 Tree diagram showing the possible genotypes of the offspring of two hybrid (Rr) parents.

A daughter inherits her father's form of this gene plus (at random) one of her mother's genes. A son inherits his father's Y chromosome (which does not carry this gene) plus (at random) one of his mother's X chromosomes (and hence one gene of this type).

There is again **complete dominance**, so that only pure recessives (bb females and b_ males) are colour blind. A Bb female is called a **carrier** of colour blindness since, though she does not have the disorder herself, she may pass it on to her children.

Suppose that, in a certain generation, the possible genotypes appear in the following proportions (for some $0 < \theta < 1$):

males	B: θ	b: $1-\theta$	
females	BB: θ^2	Bb: $2\theta(1-\theta)$	bb: $(1-\theta)^2$

A simple probability argument (see below) shows that the genotypes will reappear in the same proportions in the next generation. This is called **Hardy–Weinberg equilibrium**. Historically, the British population has been in equilibrium, with $\theta = 0.95$, giving rise to the observed proportions of 0.05 colour-blind males and just $(0.05)^2 = 0.0025$ colour-blind females.

Figure 7.5 shows the pattern of inheritance of this gene for daughters of the generation we were told about. It assumes **random mating**, that is that human beings mate independently of their (B, b) genotype.

Again, the probability associated with each 'path' through this tree diagram is found by multiplying together all the (unconditional and conditional) probabilities on the path, a simple extension of what we did before. So:

$$P(\text{daughter is BB}) = \theta^3 + \theta^2(1-\theta) = \theta^2[\theta + (1-\theta)] = \theta^2;$$

$$P(\text{daughter is Bb}) = 2\theta^2(1-\theta) + 2\theta(1-\theta)^2 = 2\theta(1-\theta)[\theta + (1-\theta)] = 2\theta(1-\theta);$$

$$P(\text{daughter is bb}) = \theta(1-\theta)^2 + (1-\theta)^3 = (1-\theta)^2[\theta + (1-\theta)] = (1-\theta)^2.$$

These are *the same probabilities* as for females of the parents' generation.

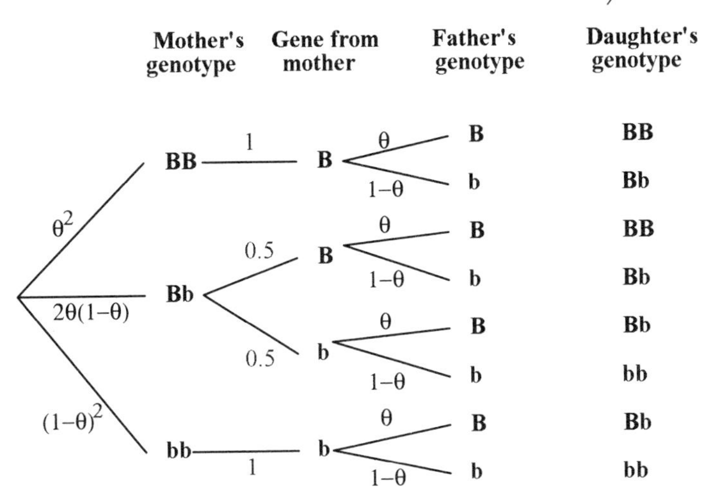

Fig 7.5 Tree diagram showing the inheritance of a sex-linked disorder.

Now prove for yourself that, among sons, the genotypes will appear in the same proportions as for males of the parents' generation. This proves that the population is in equilibrium.

For similar examples of the use of Bayes' Theorem in genetic counselling, see Murphy and Mutalik (1969).

Summary

This chapter has introduced and shown how to use the Law of Total Probability and Bayes' Theorem to combine unconditional and conditional probabilities. These extremely important results allow new information to be used to update the probability associated with an event. The updated (conditional) probability is known as the posterior probability.

FURTHER EXERCISES

1. In a multiple choice exam, four alternative answers are provided to each question, only one of which is correct. A student has probability θ $(0 < \theta < 1)$ of knowing the correct answer to a question. If she does not know the answer, she chooses one of the four alternatives at random. Find the probability that she gives the correct answer to the question. Given that she gives the correct answer to the question, find the probability that she knew the answer (and did not just guess).
2. A screening test is being devised for a particular disease. The test is very accurate: 99% of people with the disease test positive, and 95% of people who do not have the disease test negative. Assuming that 1% of those who are given the test actually have the disease, what percentage of subjects will test positive? Given that an individual tests positive, what is the posterior probability that the individual has the disease?

3. There can be problems at birth when a rhesus-negative mother gives birth to a rhesus-positive child. Of the British population, 36% are RR, 48% are Rr and 16% are rr. Assuming random mating, find out what proportion of births in the UK are potentially affected by such birth problems. [See the Application section above for more information.]
4. In the UK 5% of boys and 0.25% of girls are born with red–green colour blindness. About 51% of all the children born are boys. What proportion of children are born with red–green colour blindness? What proportion of children born with red–green colour blindness are boys? [See the Application section above for more information.]
5. (a) A woman knows that her maternal grandfather had haemophilia (an X-linked disorder), but that her other grandparents neither had the disorder themselves nor were carriers of it. Can you show that she, therefore, has probability $\frac{1}{2}$ of being a carrier of the disorder (in the absence of any other information)?
   (b) She now gives birth to her first child, a son, who does not have haemophilia. Her son's father's family has no history at all of haemophilia. Find the updated (posterior) probability that the woman is a carrier of haemophilia. [See the Application section above for more information.]
6. In a certain factory 4% of electronic components manufactured are defective. An inspector tests each component before it leaves the factory. He incorrectly rejects 2% of non-defective components and incorrectly passes 1% of defective components. What proportion of all components produced in the factory does he reject? Given that he rejects a component, what is the probability that it is not defective?

8 • Discrete Random Variables

The last few chapters have shown how to calculate and update the probabilities of events in general sample spaces. Now, we will look at what happens when we can adequately represent each outcome of an experiment by a single numerical value. A function which associates a unique real value with each outcome in the sample space is called a random variable. It is often easier to work directly with a suitable random variable than with events in the original sample space.

8.1 The probability distribution

Example 1

Suppose we decide to record the number of children born in a local maternity hospital tomorrow. A suitable sample space for this experiment is

$$S = \{0, 1, 2, \ldots\}$$

The outcome of this experiment is intrinsically numerical.

Example 2

Suppose we ask a member of the public to react to the statement 'I think the present government is managing the economy well', by ticking one of the options 'Strongly disagree', 'Disagree', 'Neutral', 'Agree', 'Strongly agree'. This is called a **Likert scale** for the response. A suitable sample space for this experiment is

$$S = \{\text{strongly disagree, disagree, neutral, agree, strongly agree}\}$$

We might, however, choose to represent these outcomes by the following numerical codes:

$$R = \{-2, -1, 0, 1, 2\}$$

In this case, though the outcomes of the experiment are not themselves numerical, they can be represented by the numbers -2, -1, 0, 1 and 2.

Example 3

Suppose we record a thermograph, a continuous trace of the ambient temperature, for 24 hours. The outcome of this experiment is a continuous, time-dependent function. Suppose that we read off the maximum temperature recorded during the day. Though the outcome of this experiment was not simply a number, we have derived a number from it.

A *function*, X, which associates a unique numerical value, $X(s)$, with every outcome $s \in S$ is called a **random variable**. Every time the experiment is conducted, one and only one value of the random variable is observed. This is called a **realization** of the random variable. It is conventional to represent a random variable by a capital letter (such as X), and a general realization of it by the corresponding lower-case letter (e.g. x).

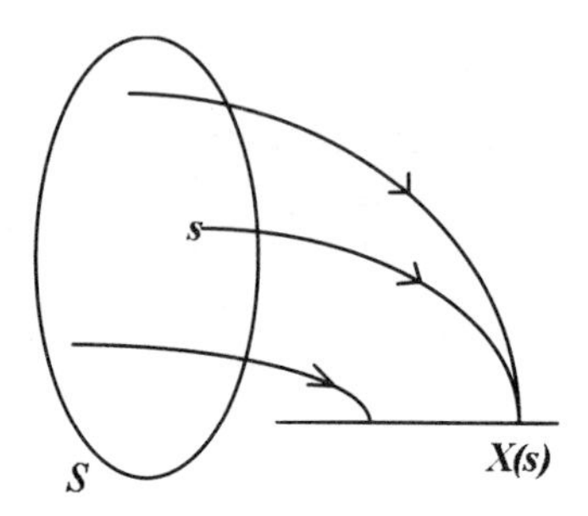

In Example 1, let the random variable X be the number of births in the hospital. Then X is the identity function

$$X(s) = s \text{ for every outcome } s \in S$$

X can take the values $0, 1, 2, \ldots$.

In Example 2, the random variable Y is defined as follows:

$$Y(\text{strongly disagree}) = -2 \qquad Y(\text{disagree}) = -1 \qquad Y(\text{neutral}) = 0$$
$$Y(\text{agree}) = 1 \qquad Y(\text{strongly agree}) = 2$$

In Example 3, the random variable Z is defined by

$$Z = \max_{0 \le t \le 24} \{g(t)\}$$

where $g(t)$ is the temperature (°C) recorded at time t.

The set of all possible values of a random variable is called its **range space**, R. In the above examples, the range spaces are:

$$R = \{0, 1, 2, \ldots\};$$
$$R = \{-2, -1, 0, 1, 2\};$$
$$R = \mathbf{R}, \text{ the set of real numbers.}$$

In Examples 1 and 2, the possible values for the random variable X, that is the values in the range space R, can be listed in the general form $x_1, x_2, \ldots$. In other words, R is a countable set (Section 2.4). We say that X is a **discrete random variable**. The range space for a discrete random variable may contain a finite or infinite number of possible values.

In Example 3, X is a measurement and can take any real value in an interval on the real line. It is clearly not possible to *list* all the possible values of X; in other words, R is uncountable. X is said to be a **continuous random variable**. Every continuous random variable is capable of assuming an infinite number of possible values.

For the next few chapters, we will deal only with discrete random variables; continuous random variables next appear in Chapter 12.

The realization of a random variable is completely determined by the outcome of the underlying experiment. Though we speak of probabilities associated with X, for example $P(X = x)$ or $P(X < x)$, these probabilities must be determined from the original sample space. For example, when X is a discrete random variable, $P(X = x)$ is just the probability of the equivalent event $\{s \in S : X(s) = x\}$. This event need not consist of just one outcome as the following example shows.

Example 4

A psychologist who is conducting learning experiments with rats has prepared a simple maze with just two narrow paths through it. When the rat is introduced into the maze, it must go down one of the paths. One leads to food, the other does not.

Each rat is introduced to the same maze three times, and the outcome on each trial is noted. A suitable sample space for this experiment is

$$S = \{\text{SSS, SSF, SFS, SFF, FSS, FSF, FFS, FFF}\}$$

where, for example, SSF means that the rat succeeds in reaching the food on the first two attempts, but fails on the third. Assuming that the rat's attempts to reach food are independent, then each of these eight outcomes is equally likely with probability $\frac{1}{8} = 0.125$.

If the random variable X is the number of times the rat succeeds in reaching the food, then:

$$P(X = 0) = P(\text{FFF}) = 0.125;$$
$$P(X = 1) = P(\text{SFF, FSF, FFS}) = 0.375;$$
$$P(X = 2) = P(\text{SSF, SFS, FSS}) = 0.375;$$
$$P(X = 3) = P(\text{SSS}) = 0.125.$$

This model does not seem very realistic. We would expect a rat to learn where the food is in the course of an experiment. Suppose that the psychologist believes that a rat has probability 0.5 of reaching the food on its first attempt, but that, each time the rat succeeds in reaching food, its chance of failing to reach food on a subsequent trial is halved. If the rat fails to reach food on a particular trial, then its chance of reaching food is unchanged for future trials. The tree diagram in Fig 8.1 indicates the probabilities associated with a rat's three attempts.

Now:

$$P(X = 0) = P(\text{FFF}) = 0.125;$$
$$P(X = 1) = P(\text{SFF, FSF, FFS}) = 0.03125 + 0.0625 + 0.125 = 0.21875;$$
$$P(X = 2) = P(\text{SSF, SFS, FSS}) = 0.046875 + 0.09375 + 0.1875 = 0.328125;$$
$$P(X = 3) = P(\text{SSS}) = 0.328125.$$

Notice that, in both models, the four probabilities add up to 1. This must be true, for the events $\{s \in S : X(s) = 0\}$, $\{s \in S : X(s) = 1\}$, $\{s \in S : X(s) = 2\}$ and $\{s \in S : X(s) = 3\}$ must *partition* the sample space, S (see Chapter 7).

First attempt	Second attempt	Third attempt	Number of successes
0.5 S	0.75 S	0.875 S	3
		0.125 F	2
	0.25 F	0.75 S	2
		0.25 F	1
0.5 F	0.5 S	0.75 S	2
		0.25 F	1
	0.5 F	0.5 S	1
		0.5 F	0

Fig 8.1 Tree diagram showing the probabilities associated with the rat learning experiment in Example 4.

Suppose that X is a discrete random variable with range space R. For any real value, x, define $p(x) = P(X = x)$. Clearly, $p(x) = 0$ when $x \notin R$. The set of pairs $\{(x, p(x)), x \in R\}$ is called the **probability distribution (p.d.)** of the random variable X. (There are many alternative names for $p(x)$, the most common being the probability function, the point probability function and the probability mass function.) The probability distribution of a discrete random variable must satisfy the following conditions:

(a) $0 \leq p(x) \leq 1$ for any real value, x;
(b) $\sum_{x \in R} p(x) = 1$.

Example 5

A small manufacturing firm employs eight production workers, of whom five are women. During a recession, the firm lays off three of its production workers, all of whom are women. If the workers to be laid off were chosen at random, find the probability distribution of X, the number of women who would be laid off. Do you think the three women actually laid off could reasonably claim that they were discriminated against on the grounds of their sex?

SOLUTION
There are $\binom{8}{3} = 56$ different ways in which to choose three workers to lay off.

X, the number of women among the three workers laid off, may take the values 0, 1, 2 and 3. The number of different ways in which to choose x women and $(3 - x)$ men is

$$\binom{5}{x}\binom{3}{3-x}$$

(see Chapter 4). So, assuming that the workers to be laid off were chosen from the workforce at random, then

$$p(x) = P(X = x) = \binom{5}{x}\binom{3}{3-x}\bigg/\binom{8}{3}$$

Using this formula, we can draw up the following table of the probability distribution of X:

x	0	1	2	3
$p(x) = P(X = x)$	$\frac{1}{56}$	$\frac{15}{56}$	$\frac{30}{56}$	$\frac{10}{56}$

Since there is probability $10/56 = 0.179$ that three workers chosen at random would all have been women, the sex-discrimination case does not seem very strong. Of course, we have not considered other relevant factors like length of service.

The random variable in Example 5 is of a special kind, known as a **hypergeometric** random variable (see Chapter 10).

Once we have defined a random variable to summarize the outcome of an experiment, we usually assume that all the interesting probabilistic information is summarized in the probability distribution. We no longer need to consider explicitly the original sample space or events within it.

EXERCISES ON 8.1

1. Find the range space of the discrete random variable, X, in each of the following examples.

(a) In a football cup tie, the result is usually decided in the standard 90 minutes of play. If the scores are level after this time, then 30 minutes of extra time are played. If the scores are still tied after extra time, then a replay is required. Again this match usually lasts 90 minutes, but 30 minutes of extra time are played if the scores are still tied at the end of normal time. If the scores are still tied after extra time in the replay, the tie is decided by tossing a coin. X is the total playing time required to settle the tie.

(b) A questionnaire asks students five questions about a lecture course they have just attended. Each of the questions requires the student to answer on a five-point Likert scale. A student's responses are scored as suggested in Example 2, and X is the total of the scores from the five questions.

(c) A multiple choice examination consists of ten questions. One mark is awarded for each correct answer, and half a mark is deducted for each wrong or missing answer. X is the total number of marks awarded to a candidate.

2. The discrete random variable X has range space $R = \{0, 1, 2, 3\}$. Which (if any) of the following functions are valid probability distributions for X?

x	0	1	2	3
$p_1(x)$	0.3	0.3	0.2	0.1
$p_2(x)$	0.3	0.3	0.2	0.2
$p_3(x)$	0.3	0.3	0.3	0.2
$p_4(x)$	0.4	0.4	0.3	-0.1

3. Consider the experiment of rolling two fair dice. Let the random variable X be the total score on the two dice. By writing out the 36 equally likely outcomes of the experiment and the value of X associated with each outcome, derive the probability distribution of X. Remember to check that the probabilities sum to 1.
4. An elderly friend of mine makes many journeys by public transport. She makes a point of complaining when the bus or train she travels on runs late or when she cannot get a seat. Her records suggest that she arrives at her destination late after 15% of her journeys; fails to find a seat on 10% of her journeys; and both fails to find a seat and arrives late on 8% of her journeys. Find the probability distribution of X, the number of reasons for complaint my friend has following one of her journeys.

8.2 The cumulative distribution function

Example 6

To test the shear strength of a batch of plastic, a rod made from the plastic is hit repeatedly with a known force until it breaks. Let the random variable X be the number of hits required until the rod breaks. If the plastic has been manufactured as specified, then there is probability 0.5 that the rod breaks on any given hit. It is assumed that the outcomes of different hits are (stochastically) independent.

The range space of X is $R = \{1, 2, 3, \ldots\}$. Also:

$$p(1) = P(X = 1) = P(\text{rod breaks on first hit}) = 0.5;$$

$$\begin{aligned} p(2) = P(X = 2) &= P(\text{rod does not break on first hit but breaks on second hit}) \\ &= P(\text{rod does not break on first hit}) \times P(\text{rod breaks on second hit}) \\ &= 0.5 \times 0.5 = 0.25; \end{aligned}$$

$$\begin{aligned} p(3) = P(X = 3) &= P(\text{rod does not break on first hit}) \\ &\quad \times P(\text{rod does not break on second hit}) \\ &\quad \times P(\text{rod breaks on third hit}) \\ &= 0.5 \times 0.5 \times 0.5 = 0.125; \end{aligned}$$

etc.

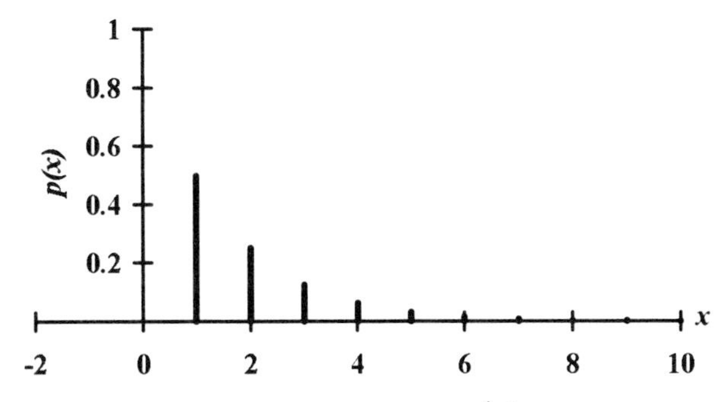

Fig 8.2 The probability distribution, $p(x) = (\frac{1}{2})^x$ $(x = 1, 2, \ldots)$, in Example 6.

In general, $p(x) = P(X = x) = (\frac{1}{2})^x$, $x = 1, 2, 3, \ldots$. This probability distribution is plotted in Fig 8.2. Though there is a non-zero probability that $X = 6, 7, \ldots$, these probabilities are too small to show up on this figure.

Notice once again that

$$\sum_{x \in R} p(x) = \sum_{x=1}^{\infty} p(x) = \frac{1}{2} + \frac{1}{4} + \frac{1}{8} + \ldots = 1$$

This is an example of a **geometric series**, which can be written in the general form

$$S_\infty = a + ar + ar^2 + ar^3 + \ldots$$

Each term of this series is found from the previous one by multiplying it by the constant r. When $-1 < r < 1$, the sum to infinity of the series, S_∞, can be found by noticing that

$$rS_\infty = ar + ar^2 + ar^3 + \ldots$$

Therefore

$$S_\infty - rS_\infty = a$$

that is

$$(1 - r)S_\infty = a$$

or

$$S_\infty = \frac{a}{1 - r}$$

When $r \geq 1$ or $r \leq -1$, then S_∞ must be infinite and the above formula does not hold.

In the above example, the series we required to sum was a geometric series with $a = \frac{1}{2}$ and $r = \frac{1}{2}$. The sum to infinity of this series is

$$S_\infty = \frac{0.5}{1 - 0.5} = 1$$

as stated.

Suppose that, in Example 6, we wanted to know the probability that three or fewer blows would be required to break the rod. We could find this sort of probability easily from the probability distribution:

$$\begin{aligned} P(X \leq 3) &= P(X = 1) + P(X = 2) + P(X = 3) \\ &= p(1) + p(2) + p(3) \\ &= 0.5 + 0.25 + 0.125 \\ &= 0.875 \end{aligned}$$

Sometimes, it is really information about $P(X \leq x)$ that is of most interest to us. The **cumulative distribution function (c.d.f.)** of a discrete random variable is defined as

$$F(x) = P(X \leq x) = \sum_{y \in R: y \leq x} p(y)$$

The cumulative distribution function is sometimes called just the distribution function. This function is defined for every real value, x, whether or not $x \in R$. Notice that:

(a) $F(-\infty) = P(X \leq -\infty) = 0$ and $F(\infty) = P(X \leq \infty) = 1$;
(b) if $x_1 \leq x_2$, then $P(X \leq x_1) \leq P(X \leq x_2)$, that is $F(x_1) \leq F(x_2)$.

In Example 6, we can readily calculate and tabulate the values of the probability distribution, and the cumulative distribution function, at the first few points in R:

x	1	2	3	4	...
$p(x) = P(X = x)$	0.5	0.25	0.125	0.0625	...
$F(x) = P(X \leq x)$	0.5	0.75	0.875	0.9375	...

A graph of this cumulative distribution function is shown in Fig 8.3. Since X is a discrete random variable with range space $\{1, 2, 3, \ldots\}$, then the value of $F(x)$ only changes at $x = 1, 2, 3, \ldots$.

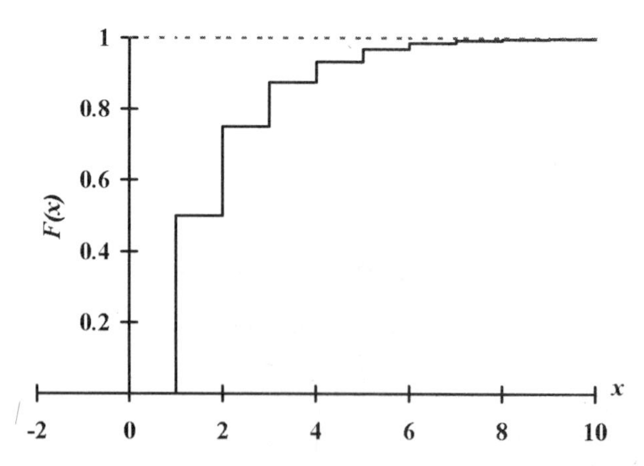

Fig 8.3 The cumulative distribution function of the discrete random variable in Example 6.

EXERCISES ON 8.2

1. Plot the probability distributions and cumulative distribution functions of the random variables introduced in Exercises 3 and 4 on p. 62.
2. When X is the random variable introduced in Example 6 on p. 62, show that the probability that X is an even number is $\frac{1}{3}$. [You will need to use the formula for the sum to infinity of a geometric series.]

Summary

This chapter has introduced the concept of a random variable, which is a function that associates a unique real value with every outcome in a sample space. The set of all possible values of the random variable is called its range space. A discrete random variable has a range space whose elements can be listed in the form $x_1, x_2, \ldots$; there may be either a finite or an infinite number of these values. The sets of points $\{(x, p(x)), x \in R\}$, where $p(x) = P(X = x)$, is called the probability distribution of the discrete random variable X. The corresponding cumulative distribution function is defined (for all real values, x) by $F(x) = P(X \leq x)$. Probabilities associated with a random variable must be determined from events in the original sample space.

FURTHER EXERCISES

1. The discrete random variable X has range space $R = \{-1, 0, 1\}$. The probabilities that X takes these values are, respectively, $[(1-\theta)/2], \theta, [(1-\theta)/2]$, where θ is a real constant. Show that, for this to be a valid probability distribution function, θ must lie in the range 0 to 1.
2. The discrete random variable X has range space $R = \{-3, -1, 1, 3\}$. The probabilities that X takes these values are, respectively, $[(1-\theta)/4]$, $[(1-3\theta)/4]$, $[(1+3\theta)/4]$ and $[(1+\theta)/4]$, where θ is a real constant. Find the range of values of θ for which this is a valid probability distribution.
3. Let X be the number of children in a completed family in a particular society. It has been suggested that X has a probability distribution of the form

$$p(0) = \lambda$$
$$p(x) = k\theta^{x-1} \qquad x = 1, 2, \ldots$$

   where $0 < \lambda < 1$ and $0 < \theta < 1$ and k is a real constant. Show that the conditions for a valid probability distribution require that $k = (1-\lambda)(1-\theta)$.
4. A drunk man requires to make his way along a straight road to reach home. Each time he moves, he has probability θ of stepping forward 1 metre and probability $1-\theta$ of stepping back 1 metre, independently of what happens every other time he moves. Let X_t denote the overall distance forward he has moved after t steps. Derive the probability distribution and the cumulative distribution function of X_t when (a) $t = 3$; (b) $t = 4$.
5. Consider the sum of the first n terms of the general geometric series:

$$S_n = a + ar + ar^2 + \ldots + ar^{n-1}$$

Using a similar argument to that used to derive the sum to infinity of this series, show that

$$S_n = \frac{a(1 - r^n)}{(1 - r)} \qquad (r \neq 1)$$

Use this result to show that the cumulative distribution function of the random variable in Example 6 takes the values $F(x) = 1 - (\frac{1}{2})^x$ $(x = 1, 2, \ldots)$.

9 • The Binomial and Related Distributions

There are a small number of families of discrete probability distributions that have been found to be particularly useful. They accurately represent the critical features shared by many experiments in many application areas. This chapter introduces the Binomial distribution, one of the simplest and most widely used models of all. The Geometric distribution, which is related to the Binomial distribution, is also introduced.

9.1 The Binomial distribution

Here are brief descriptions of random variables associated with four experiments. Though they arise in very different contexts, they share a number of crucial features.

(1) X is the number of heads recorded in 1000 tosses of a fair coin.
(2) X is the number of times a score of 6 is recorded in 100 rolls of a fair die.
(3) Twenty electronic components produced consecutively on a production line are tested; X is the number that are found to be satisfactory.
(4) X is the number of girls in a family of four children.

In each case, a basic experiment is replicated a number of times (tossing a coin, rolling a die, testing a component, recording the sex of a child). We call the individual replicates **trials**. The sequences of trials described above share four common features:

(a) Each trial has *two possible outcomes*, considered as a success and a failure. In the above examples, success is equated with obtaining heads, a score of 6, a satisfactory result to the test and a girl.
(b) The outcomes of the trials are *stochastically independent*.
(c) In each trial, the probability of a success has the *same value*, θ (where $0 \leq \theta \leq 1$ and θ is often unknown).
(d) The number of trials is *fixed in advance* ($n = 1000, 100, 20, 4$).

Trials with the features (a) to (c) are called **Bernoulli trials**. When X is the number of successes in n Bernoulli trials, each with success probability θ, then X is said to be a **binomial** random variable. This is written $X \sim \text{Bi}(n, \theta)$.

So, in the examples above, the random variables have the following distributions:

(1) $\text{Bi}(1000, 0.5)$;
(2) $\text{Bi}(100, 1/6)$;

(3) $\text{Bi}(20, \theta)$ where θ is the probability that a randomly selected component is satisfactory;
(4) $\text{Bi}(4, \theta)$ where θ is the probability that a randomly selected child is a girl (does $\theta = 0.5$?).

Notice particularly that these are only binomial random variables if property (b) – independence – holds. Can you think of reasons why this might not be true in some of experiments (1) to (4)?

Suppose that $X \sim \text{Bi}(n, \theta)$. Then the range space of X is $R = \{0, 1, \ldots, n\}$, since there are n trials, each of which may result in a success or a failure. For any $x \in R$, the following is a typical sequence of results with successes on x *given* trials:

$$\underbrace{\text{S S} \ldots \text{S}}_{x} \; \underbrace{\text{F F} \ldots \text{F}}_{n-x}$$

Since each success occurs with probability θ, each failure occurs with probability $1 - \theta$, and the trials are independent, it follows that the probability of this *given* sequence of results is

$$\theta \cdot \theta \cdot \ldots \cdot \theta \cdot (1-\theta) \cdot (1-\theta) \cdot \ldots \cdot (1-\theta) = \theta^x (1-\theta)^{n-x}$$

There are $\binom{n}{x}$ different sequences with exactly x successes, and each of these sequences has the same probability. So, X has probability distribution

$$p(x) = P(X = x) = \binom{n}{x} \theta^x (1-\theta)^{n-x} \quad (x = 0, 1, \ldots, n)$$

This is a valid probability distribution, since:

(a) $0 \leq p(x) \leq 1$ for all x;
(b)

$$\begin{aligned} \sum_{x=0}^{n} p(x) &= \sum_{x=0}^{n} \binom{n}{x} \cdot \theta^x \cdot (1-\theta)^{n-x} \\ &= [\theta + (1-\theta)]^n \quad \text{(Binomial Theorem)} \\ &= 1 \end{aligned}$$

Example I

In the course of a printing process, 20 different colours are inked onto a sheet of Perspex. The colours are inked on consecutively and independently, 2.5% of items being inked unsatisfactorily with each colour. Find the overall wastage rate for this process.

SOLUTION

Let X be the number of colours inked unsatisfactorily onto a randomly selected sheet of Perspex. Then, $X \sim \text{Bi}(20, 0.025)$. So,

$$
\begin{aligned}
P(\text{an item is unsatisfactory}) = P(X > 0) &= 1 - P(X = 0) \\
&= 1 - \binom{20}{0}(0.025)^0(1-0.025)^{20} \\
&= 1 - (0.975)^{20} \\
&= 0.397
\end{aligned}
$$

So, though the wastage rate at each stage is very low (just 2.5%), overall almost 40% of production is wasted. These figures are well known to the management of a printing factory I visited recently.

Example 2

Suppose that $X \sim \text{Bi}(10, 0.1)$. Write out the probability distribution of X.

SOLUTION

In this case it is easy to obtain the probability distribution directly from the definition. We will, however, use the following recursive formula, which is particularly useful when carrying out hand calculations or programming a calculator or computer when n is large or θ is small.

Suppose that $X \sim \text{Bi}(n, \theta)$. Then, for $x = 0, 1, \ldots, n-1$,

$$
\begin{aligned}
p(x+1) &= \binom{n}{x+1}\theta^{x+1}(1-\theta)^{n-(x+1)} \\
&= \frac{n!}{(x+1)!(n-x-1)!}\theta^{x+1}(1-\theta)^{n-x-1} \\
&= \frac{n!}{x!(n-x)!}\theta^{x}(1-\theta)^{n-x}\left(\frac{n-x}{x+1}\cdot\frac{\theta}{1-\theta}\right) \\
&= p(x)\left(\frac{n-x}{x+1}\cdot\frac{\theta}{1-\theta}\right)
\end{aligned}
$$

For Example 2, then, we can work out the probability distribution of X as follows:

$$p(0) = \binom{10}{0}(0.1)^0(0.9)^{10} = (0.9)^{10} = 0.349;$$

$$p(1) = p(0)\cdot\frac{10}{1}\cdot\frac{0.1}{0.9} = 0.387;$$

$$p(2) = p(1)\cdot\frac{9}{2}\cdot\frac{0.1}{0.9} = 0.194;$$

$$p(3) = p(2)\cdot\frac{8}{3}\cdot\frac{0.1}{0.9} = 0.057;$$

$$p(4) = p(3)\cdot\frac{7}{4}\cdot\frac{0.1}{0.9} = 0.011;$$

$$p(5) = p(4)\cdot\frac{6}{5}\cdot\frac{0.1}{0.9} = 0.001;$$

etc.

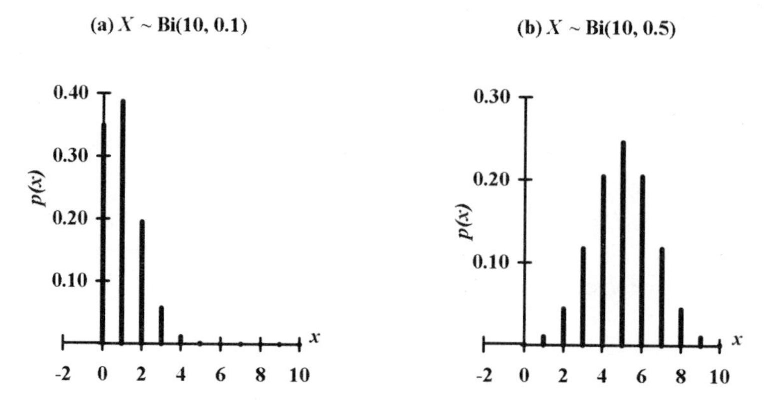

Fig 9.1 The probability distribution of X when (a) $X \sim \text{Bi}(10, 0.1)$; (b) $X \sim \text{Bi}(10, 0.5)$.

This probability distribution is plotted in Fig 9.1(a). Figure 9.1(b) shows the probability distribution of X when X has another Binomial distribution, $X \sim \text{Bi}(10, 0.5)$. Whenever $\theta = 0.5$, the probability distribution of a $\text{Bi}(n, \theta)$ random variable is symmetric: $p(x) = p(n - x)$. Otherwise, the probability is skew symmetric. (See also Tutorial Problem 1.)

Example 3

An experiment consists of subjecting an individual to a series of three different stimuli and recording each time whether or not the subject responds. The probability of a response to each stimulus is θ, and a subject is believed to respond independently to the three stimuli. Find the probability that:

(a) a series of three stimuli causes at least one response;
(b) of five consecutive series of stimuli, at most one consists entirely of responses.

SOLUTION

(a) Let X denote the number of responses caused by a series of three stimuli. Then, assuming that responses to the different stimuli are independent, $X \sim \text{Bi}(3, \theta)$. So, $P(X \geq 1) = 1 - P(X = 0) = 1 - (1 - \theta)^3$.
(b) Let Y denote the number of series, out of five consecutive series, that consist entirely of responses. Assuming that the series are independent, $Y \sim \text{Bi}(5, \phi)$ where ϕ is the probability that a series consists entirely of responses.
Now, $\phi = \theta^3$. So

$$P(Y \leq 1) = \binom{5}{0}\phi^0(1 - \phi)^5 + \binom{5}{1}\phi^1(1 - \phi)^4 = (1 - \theta^3)^5 + 5\theta^3(1 - \theta^3)^4$$

EXERCISES ON 9.1

1. Suppose that $X \sim \text{Bi}(4, 0.25)$. Write out the probability distribution of X. Hence write out the values of the cumulative distribution function, $F(x)$, at $x = 0, 1, 2, 3$ and 4.

2. Suppose that $Y \sim \text{Bi}(4, 0.75)$. Write out $p(y)$ and $F(y)$ $(y = 0, 1, 2, 3, 4)$. Write down any relationships you can spot between the probability distribution and the cumulative distribution function of Y and those of X in the exercise above.
3. Now try to generalize the results of Exercise 2 above as follows. Suppose that $X \sim \text{Bi}(n, \theta)$ and $Y \sim \text{Bi}(n, 1 - \theta)$. Then, if X is the number of successes in n Bernoulli trials, Y can be thought of as the number of failures. Show that, for $x = 0, 1, \ldots, n$, (a) $P(X = x) = P(Y = n - x)$ and (b) $P(X \leq x) = P(Y \geq n - x)$.
4. A recent survey has shown that 20% of pre-school children in the UK have speech or hearing difficulties. Write down a formula for $p(x) = P(X = x)$, when X is the number of children in a nursery class of 20 children who have speech or hearing difficulties. Staff claim that, if the class were to contain more than two children with speech or hearing difficulties, then the work of the class could be adversely affected; find the probability of this event.
5. Newly diagnosed sufferers from hay fever are being recruited into the clinical trial of a new drug to relieve their symptoms. Each subject is randomly allocated to be treated either with the new drug or with a drug that is currently commonly prescribed (with probability 0.5 of receiving each treatment). If six subjects are recruited from a particular general practitioner's list, find the probability that all six receive the same drug treatment.

9.2 The Geometric distribution

Here are variations of the experiments (1)–(3) listed at the start of this chapter.

(1′) A fair coin is tossed repeatedly, until it lands tails up for the first time.
(2′) A fair die is rolled repeatedly, until the first score of 1, 2, 3, 4 or 5 is recorded.
(3′) Components off a production line are tested, in turn, until the first defective item is found.

The common features of these experiments can be expressed in the following way.

(a) There is a *potentially infinite* series of Bernoulli trials, each with 'success' probability θ $(0 < \theta < 1)$.
(b) The trials continue until *the first failure* occurs. (Here, failure is equated with tails, a score of less than 6, a defective component.)

Let the random variable X be the total number of trials required until the first failure occurs. Then X is a **geometric** random variable with parameter θ, written $X \sim \text{Ge}(\theta)$.

In the examples above: (1′) $X \sim \text{Ge}(0.5)$; (2′) $X \sim \text{Ge}(1/6)$; (3′) $X \sim \text{Ge}(\theta)$, where θ is the probability that a randomly selected component is *satisfactory*. It is sometimes preferable to think of a $\text{Ge}(\theta)$ random variable as the number of Bernoulli trials until the first 'success', when $P(\text{'success'}) = 1 - \theta$ (or $P(\text{'failure'}) = \theta$).

Suppose that $X \sim \text{Ge}(\theta)$. Then the range space of X is $R = \{1, 2, \ldots\}$. For any $x \in R$,

$$P(X = x) = P\begin{pmatrix}\text{'success'}\\ \text{on first trial}\end{pmatrix} \times \ldots \times P\begin{pmatrix}\text{'success'}\\ \text{on } (x-1)\text{th trial}\end{pmatrix} \times P\begin{pmatrix}\text{'failure'}\\ \text{on } x\text{th trial}\end{pmatrix} = \theta^{x-1}(1-\theta)$$

So X has probability distribution

$$p(x) = P(X = x) = \theta^{x-1}(1-\theta) \quad (x = 1, 2, \ldots)$$

The terms in the probability distribution form a *geometric series*, with first term $(1-\theta)$ and constant ratio θ, where $0 \leq \theta \leq 1$ (see Example 6 in Chapter 8).

This is a valid probability distribution since:

(a) $0 \leq p(x) \leq 1$, for all x;
(b)

$$\begin{aligned}\sum_{x=1}^{\infty} p(x) &= \sum_{x=1}^{\infty} \theta^{x-1}(1-\theta)\\ &= (1-\theta) \times (1 + \theta + \ldots)\\ &= (1-\theta) \times \frac{1}{1-\theta} \quad \text{(sum of geometric series)}\\ &= 1\end{aligned}$$

Example 4

In BBC TV's programme *Call My Bluff*, contestants are presented with an uncommon English word and three alternative meanings for it, only one of which is correct. The contestant gets one chance to identify the correct meaning of each word. Supposing that a particular contestant guessed the correct meaning on each of his or her turns, find the probability that the contestant would need three or fewer turns to record the first correct answer.

SOLUTION

Let X be the total number of turns a contestant requires to record the first correct answer. Since the contestant guesses each time, then the turns must be a sequence of Bernoulli trials, with constant probability $\theta = 2/3$ of giving a wrong answer. So, $X \sim \text{Ge}(2/3)$, that is

$$p(x) = \left(\frac{2}{3}\right)^{x-1}\left(\frac{1}{3}\right) \quad x = 1, 2, 3, \ldots$$

We require to find

$$P(X \leq 3) = \frac{1}{3} + \frac{2}{3}\cdot\frac{1}{3} + \frac{2}{3}\cdot\frac{2}{3}\cdot\frac{1}{3} = \frac{19}{27} = 0.704$$

EXERCISES ON 9.2

1. Suppose that $X \sim \text{Ge}(0.2)$. Find the probability that $X = 1$, $X = 2$, $X \geq 3$, $X \leq 3$.
2. Use the formula for the sum to n terms of a geometric series (Chapter 8) to find the cumulative distribution function of the $\text{Ge}(\theta)$ distribution. Plot this function when (a) $\theta = 0.2$; (b) $\theta = 0.8$.
3. People with chronic diseases (such as asthma) sometimes have acute attacks and must be admitted to hospital to have their symptoms brought back under control. A single hospital visit does not always succeed in controlling the symptoms, with the result that the patient must be readmitted soon after being discharged. Suppose that X, the total number of admissions required to control the symptoms in a randomly selected patient, has a $\text{Ge}(0.15)$ distribution. Find the probability that a patient requires to be readmitted (a) at least once, (b) at least twice, after the initial admission to hospital. How realistic do you find a geometric model in this example?
4. As a generalization of the Geometric distribution, let the random variable X be the total number of Bernoulli trials required until the kth failure occurs $(k \geq 1)$. X is said to have a **Negative Binomial** (or **Pascal**) distribution with parameters k and θ, written $X \sim \text{NeBi}(k, \theta)$. Show that the probability distribution of X is

$$p(x) = P(X = x) = \binom{x-1}{k-1}\theta^{x-k}(1-\theta)^k \quad (x = k, k+1, \ldots)$$

Application: winning strategies in racquet sports

In some racquet sports (e.g. squash, badminton, racquetball), only one of the two players (the current server) may score at each stage. Service alternates between the players in the course of a game, according to the following rules. If the server wins a rally, he or she scores a point and continues to serve. If the other player (the receiver of the service) wins the rally, neither player wins a point but service passes to the receiver.

Suppose that player S has probability θ of winning a *rally* on his own service when playing against player T. Suppose also that S has probability $(1 - \phi)$ of winning a *rally* on T's service. These are simplifying assumptions; in practice, the probability of winning a rally will fluctuate in the course of a game. We will now find the probability that S wins the first *point* when he serves at the start of play. Table 9.1 shows some of the possible sequences of rallies required to decide the winner of the first point. The point is won only when one of the players wins a rally when serving. This means that S, who is serving at the start, can only win the point on an odd-numbered rally. T can only win on an even-numbered rally. For example, in outcome 2, S serves first but loses the first rally. So T, having won the first rally, serves for the second rally which he wins and so wins the point.

Table 9.1 Some possible sequences of rallies to decide the first point.

Outcome		Rally no. 1	2	3	4	5	...	Winner of point
1	Server for rally	S						S
	Winner of rally	S						
2	Server for rally	S	T					T
	Winner of rally	T	T					
3	Server for rally	S	T	S				S
	Winner of rally	T	S	S				
4	Server for rally	S	T	S	T			T
	Winner of rally	T	S	T	T			
5	Server for rally	S	T	S	T	S		S
	Winner of rally	T	S	T	S	S		
...	...	...	...	...	...	...	...	...

Assuming that all the rallies are won and lost independently of one another (another crucial assumption that will not be strictly true in practice), then

$$\begin{aligned}&P(\text{S wins the first point on the first rally} \mid \text{S serves at the start of play})\\&\quad = P(\text{S wins the first rally on his service}) = \theta;\end{aligned}$$

$$\begin{aligned}&P(\text{S wins the first point on the third rally} \mid \text{S serves at the start of play})\\&\quad = P(\text{S loses the first rally on his service}) \times P(\text{S wins the second rally}\\&\qquad \text{on T's service}) \times P(\text{S loses the third rally on his service})\\&\quad = (1-\theta)(1-\phi)\theta;\end{aligned}$$

$$\begin{aligned}&P(\text{S wins the first point on the fifth rally} \mid \text{S serves at the start of play})\\&\quad = P(\text{S loses the first rally on his service}) \times P(\text{S wins the second rally}\\&\qquad \text{on T's service}) \times P(\text{S wins the third rally on his service})\\&\qquad \times P(\text{S wins the fourth rally on T's service})\\&\qquad \times P(\text{S wins the fifth rally on his service})\\&\quad = (1-\theta)(1-\phi)(1-\theta)(1-\phi)\theta;\end{aligned}$$

etc. Hence

$$\begin{aligned}&P(\text{S wins the first point} \mid \text{S serves at the start of play})\\&\quad = \theta + [(1-\theta)(1-\phi)]\theta + [(1-\theta)(1-\phi)]^2\theta + \ldots \quad \text{[a geometric series]}\\&\quad = \frac{\theta}{1-(1-\theta)(1-\phi)} = \alpha \text{ (say)}\end{aligned}$$

In fact, whenever S serves to start the play for a new point (having won the previous point), he has probability α of winning that point, so T has probability $1-\alpha$ of winning the point. Similarly, T has probability

$$\beta = \frac{\phi}{1-(1-\theta)(1-\phi)}$$

of winning a point when she serves to start the play for that point (having won the previous point), so S wins the point with probability $1-\beta$.

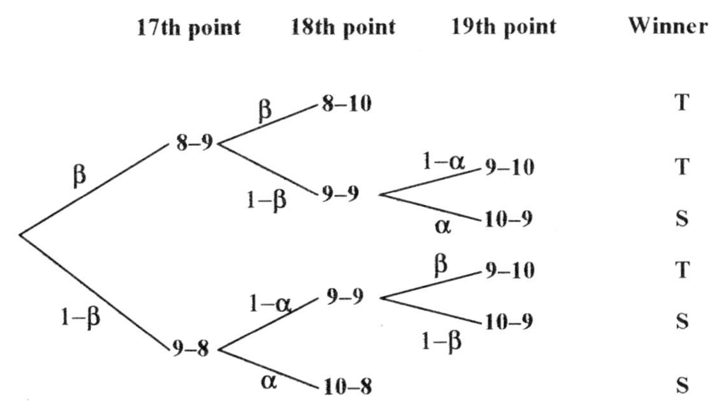

Fig 9.2 Tree diagram showing the possible sequences of points to complete a game of squash when player T is serving at 8–8.

A game of squash is usually won by the first player to score nine points. When the score reaches 8–8, the player receiving serve (i.e. the player who lost the 16th point of the game) may choose whether the game should be decided on the next point or whether it should now continue until one of the players wins ten points. Suppose player T ties the score at 8–8. Which is the better choice for S to make?

If S chooses to make the next point decisive, then the probability that he wins the game is $1 - \beta$, which is the probability that he wins the next point (with T serving). If S opts instead to play until one of them wins ten points, then his probability of winning is $(1 - \beta)(1 - \beta + 2\alpha\beta)$ (see Fig 9.2). So, S should choose to play to nine points only if

$$(1 - \beta)(1 - \beta + 2\alpha\beta) < (1 - \beta)$$

that is,

$$\alpha < 0.5$$

After some further algebra, this becomes $[\theta/(1 - \theta)] < \phi$, which gives the decision regions shown in Fig 9.3. Clearly, S should only choose to play to nine points if he is a lot stronger at receiving service than at serving.

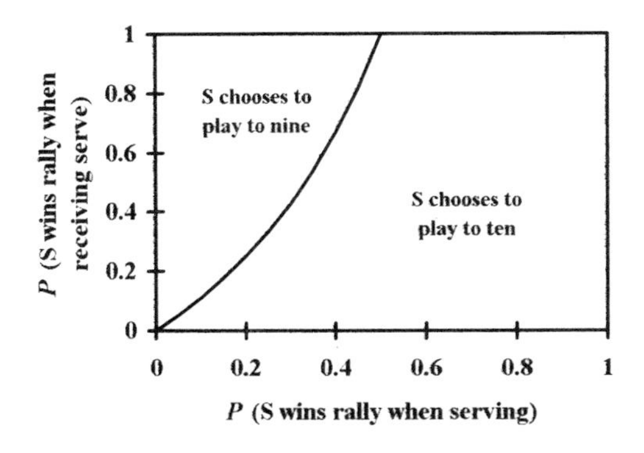

Fig 9.3 Optimal decisions at 8–8 for a player in a game of squash. After Simmons (1989).

For further information about how probability models can be applied to squash, badminton and similar sports, see Simmons (1989) and Strauss and Arnold (1987). George (1973), following Hsi and Burych (1971), uses a simple probability model as the basis for a discussion of optimal serving strategies in tennis. Reep and Benjamin (1968) and Reep *et al.* (1971) base a discussion of strategies in association football and other team games on the Geometric distribution.

Summary

This chapter has introduced the Binomial and Geometric distributions, which are commonly used to model discrete random variables in a wide variety of contexts. The correct use of these distributions is (usually) restricted to experiments that consist of a sequence of Bernoulli trials. The number of successes recorded in a fixed number of Bernoulli trials is a binomial random variable. The total number of Bernoulli trials required to record the first failure follows a Geometric distribution. Some applications of probability to determine optimal strategies in sport have also been introduced.

TUTORIAL PROBLEM 1

For a given value of n (say, $n = 10$ or 20), investigate the shape of the $\mathrm{Bi}(n, \theta)$ distribution for various values of θ (say, $\theta = 0.05, 0.1, 0.2, \ldots,$ $0.9, 0.95$). This is most easily done using a computer spreadsheet, programmed either with the formula for $p(x)$ directly or using the recursion formula introduced in Example 2. Repeat the exercise for a given θ and various values of n. Describe your results.

TUTORIAL PROBLEM 2

Using the references given in the Application section above, which you should be able to obtain through your lecturer, tutor or library, investigate some of the ways in which probability models have been applied to a sport that you play. Do you think these models are based on sensible assumptions? How much new insight have these models given you?

FURTHER EXERCISES

1. Sequences of zeros and ones are transmitted along a binary data channel. Owing to the presence of noise on the channel, each digit independently may be wrongly received, that is a 0 may be sent but a 1 received or vice versa.
   (a) The probability that a digit is wrongly received is $\theta = 0.001$. Find the probability that, in a message of 10,000 digits, at least one digit is wrongly received.

(b) In an effort to reduce even further the proportion of wrongly received digits, each digit is transmitted in triplicate. When a triple is scanned at the receiving end, the digit recorded is the one which occurs more frequently in the triple. What is the probability that a digit sent in triplicate is wrongly received? Find the probability that, in a message of 10,000 digits, at least one digit is now wrongly received.

2. A triangle test is being conducted to find out whether or not consumers can tell the difference between two brands of cola. This means that a subject is presented with three apparently identical glasses of cola, two of one brand and one of the other. The subject has to identify the odd one out. This basic trial is repeated five times in all with each subject. A guesser is someone who cannot tell the difference between the two brands of cola and guesses the answer each time.

(a) Find the probability that a guesser correctly identifies the odd one out on at least two of the five trials.

(b) Find the probability that a guesser requires at most two trials to record the first correct answer.

3. An aeroplane is fitted with four engines, but is able to make a normal landing as long as at least two of its engines continue to function normally. In a typical flight, the engines each have probability θ $(0 < \theta < 1)$ of failing, independently of one another. Find the probability that, at the end of a typical flight, the aeroplane is able to land normally. How realistic do you find the assumptions stated above?

4. In the game of table tennis, a player serves for five points in a row. Both players may win each point, no matter who is serving. Susan believes that, when serving in a game against Theresa, she has probability 0.6 of winning a point. Assuming that the outcomes of different points are independent, find the probability that Susan wins at least three out of five points on her service.

5. Three players, A, B and C, are playing a game. Each in turn rolls a fair die. The winner is the player who throws the first 6. Show that A, B and C win with respective probabilities 36/91, 30/91 and 25/91. [You should apply a similar argument to that used to derive α in the above Application section.]

6. Suppose that two players, S and T, are playing a game of tennis. S serves throughout the game, but in tennis either player can win a point, no matter who serves. S wins any rally (and, hence, any point) with probability θ $(0 < \theta < 1)$. S wins the game if either (1) S wins four rallies before T wins three, or (2) each player wins three of the first six rallies, and S is the first player subsequently to win two rallies in a row from any position where the score is tied. Assuming that rallies are won independently of one another, show that S wins the game with probability

$$\theta^4(15 - 24\theta + 10\theta^2) + \frac{20\theta^5(1-\theta)^3}{1 - 2\theta(1-\theta)}$$

Check that this probability equals 0.5 when θ equals 0.5.

10 • Other Special Discrete Distributions

In Chapter 9, we introduced the Binomial distribution, which is arguably the most useful model for discrete random variables. In this chapter, we meet the Poisson and the Hypergeometric distributions, which are both used widely. We shall also see that, although each of these is important in its own right, both are closely related to the Binomial distribution.

10.1 The Poisson distribution

The conditions that give rise to a Binomial distribution are so simple, and occur so often in practice, that Binomial distributions are the most common models adopted for discrete random variables. They do not, however, fit all situations and there are other common experimental conditions that generate other families of discrete random variables. Poisson random variables also occur sufficiently often to make this an important model to study. Here are some examples of experiments that might give rise to Poisson random variables.

(1) X is the number of emissions from a radioactive source in one minute.
(2) X is the number of telephone calls received at a particular telephone exchange in a period of one hour.
(3) X is the number of vehicles passing a fixed point on a motorway in the space of five minutes.
(4) X is the number of newsagents' shops in a 500 m $\times$ 500 m area of an urban conurbation.

In all these experiments, *isolated events* (radioactive emissions, telephone calls, traffic arrivals, sites of shops) are occurring in *continuous* time or space. Experiments that give rise to Poisson random variables share the following further features.

(a) The numbers of events that occur in non-overlapping segments of time or space are (stochastically) *independent*.
(b) Events occur *singly* rather than in groups. So, the probability that two or more events occur in a small enough time interval or spatial area is negligible.
(c) Events are occurring at a *constant average rate* per unit time or area, throughout the whole period or region of interest. To make this condition plausible, it might be necessary to restrict the experimental conditions in some way, for example to telephone calls received between 9 a.m. and 5 p.m. on weekdays.

A process that generates sequences of events with the features (a) to (c) is called a **Poisson process**. When a random variable, X, is defined to be the number of events from a Poisson process that occur in a *fixed length of time*, then X is said to be a **Poisson** random variable. This is written $X \sim \text{Po}(\theta)$, where $\theta > 0$ is the (constant) average number of events that occur in a time interval of this length.

Suppose that $X \sim \text{Po}(\theta)$. The range space of X is $R = \{0, 1, 2, \ldots\}$. It is possible to prove from the conditions (a) to (c) that X has the following probability distribution:

$$p(x) = P(X = x) = \frac{e^{-\theta}\theta^x}{x!} \quad (x = 0, 1, 2, \ldots)$$

This is a valid probability distribution since:

(a) $0 \leq p(x) \leq 1$ for any real value, x;

(b)

$$\begin{aligned}\sum_{x=0}^{\infty} p(x) &= \sum_{x=0}^{\infty} \frac{e^{-\theta}\theta^x}{x!} \\ &= e^{-\theta}\left(1 + \theta + \frac{\theta^2}{2!} + \frac{\theta^3}{3!} + \ldots\right) \\ &= e^{-\theta}e^{\theta} \quad \text{(Maclaurin expansion of } e^{\theta}) \\ &= 1\end{aligned}$$

Example 1

In one of the most famous historical applications of the Poisson distribution, it was used to model the number of deaths from horse-kicks in Prussian cavalry regiments in the years 1875–94. X, the number of deaths from horse-kicks suffered in a cavalry corps in one year, was found to be a Po(0.7) random variable. So, the probability that no one in a corps was killed by a horse-kick in one year was

$$p(0) = P(X = 0) = \frac{e^{-0.7}(0.7)^0}{0!} = \frac{e^{-0.7}1}{1} = e^{-0.7} = 0.497$$

Example 2

A recent survey (*AA Magazine*, 1994) has shown that, on average, two major repairs must be made to a house in the UK in any year. Let the random variable X be the number of major repairs required in a randomly selected house next year. It is quite plausible that $X \sim \text{Po}(2)$. We will now calculate some of the values of the probability distribution and the cumulative distribution function of X.

The following recursion formula reduces the amount of calculation involved. If $X \sim \text{Po}(\theta)$, then for $x \geq 0$,

$$p(x+1) = \frac{e^{-\theta}\theta^{x+1}}{(x+1)!} = \frac{e^{-\theta}\theta^x}{x!} \cdot \frac{\theta}{x+1} = p(x) \cdot \frac{\theta}{x+1}$$

So

$$p(0) = \frac{e^{-\theta}\theta^0}{0!} = \frac{e^{-\theta}1}{1} = e^{-\theta}$$

$$p(1) = p(0) \cdot \frac{\theta}{1}$$

$$p(2) = p(1) \cdot \frac{\theta}{2}$$

etc.

When $\theta = 2$,

$$p(0) = e^{-2} = 0.1353 \quad p(1) = p(0) \cdot \frac{2}{1} = 0.2707$$

$$p(2) = p(1) \cdot \frac{2}{2} = 0.2707 \quad \text{etc.}$$

Table 10.1 gives the first few values of the probability distribution and the cumulative distribution function of the Po(2) distribution. This shows that, for example, the probability that more than two major repairs will be required to a house next year is

$$P(X > 2) = 1 - P(X \leq 2) = 1 - F(2) = 0.3233$$

In other words, in the next calendar year, about one-third of British houses will require three or more major repairs.

Table 10.1 Some values of the p.d. and c.d.f. of the Po(2) distribution.

x	0	1	2	3	4	5	...
$p(x) = P(X = x)$	0.1353	0.2707	0.2707	0.1805	0.0902	0.0361	...
$F(x) = P(X \leq x)$	0.1353	0.4060	0.6767	0.8571	0.9473	0.9834	...

At this point it is useful to expand on the previous definition of a Poisson process. Emissions from radioactive sources have been shown to satisfy the requirements for a Poisson process. Suppose that a radioactive source is emitting α-particles at an average rate of 32 particles per minute. This means that, if the number of emissions is counted for one-eighth of a minute (7.5 seconds), then on average $32/8 = 4$ emissions will occur. So, the number of particles emitted in an interval of one-eighth of a minute is a Po(θ) random variable with $\theta = 4$.

Similarly, the average number of particles emitted in a quarter of a minute (15 seconds) is $32/4 = 8$, so the number of particles emitted in an interval of a quarter of a minute is a Po(θ) random variable with $\theta = 8$. In general, the number of particles emitted in an interval of length t minutes will be a Po($32t$) random variable.

Whenever a Poisson process is generating events at an average rate of λ events per minute, then the number of events occurring in an interval of length t minutes is a Po(λt) random variable. We will refer to Poisson processes again in Chapters 12 and 17, when we will use this characterization of a Poisson process to obtain some further important results.

The next example shows how a Poisson distribution can sometimes be used as an approximation to a Binomial distribution.

Example 3

A manufacturer sells screws in boxes of 100. Just 0.005 of all screws produced in this factory are defective. Find the probability that no more than one screw in a box is defective.

Let the random variable X be the number of defective screws in a box of 100. Then, $X \sim \text{Bi}(100, 0.005)$. So, the required probability is

$$P(X \leq 1) = p(0) + p(1) = (0.995)^{100} + 100(0.005)(0.995)^{99}$$

This is an example where n $(= 100)$ is reasonably large, θ $(= 0.005)$ is small, but the product $n\theta$ $(= 0.5)$ is moderately large. Notice that $\lambda = n\theta$ is the average number of defective screws in a box (in a sense we will make clearer in the next chapter). In these circumstances, probabilities are cumbersome to work out directly from the Binomial probability distribution. It can be shown that $X \approx \text{Po}(\lambda)$, in the sense that

$$p(x) \approx \frac{e^{-\lambda}\lambda^x}{x!} \quad (x = 0, 1, 2, \ldots)$$

It is much easier to work out probabilities from this Poisson distribution than from the original Binomial distribution. These approximate values can be very close to the correct answer. In Example 3:

$X \sim \text{Bi}(100, 0.005)$ gives the exact probability, $P(X \leq 1) = 0.9102$;

$X \approx \text{Po}(0.5)$ gives the approximate value, $P(X \leq 1) = 0.9098$.

It can be proved analytically that, if $n \to \infty$ and $\theta \to 0$ in such a way that $n\theta$ is held constant, then each term in the $\text{Bi}(n, \theta)$ probability distribution tends to the value of the equivalent term in the $\text{Po}(n\theta)$ distribution. This proof is too difficult to give here in full, but we can use the following argument to show that this is a plausible result.

Suppose that $n\theta = \lambda$. If $X \sim \text{Bi}(n, \theta)$, then for $x = 0, \ldots, n-1$,

$$\begin{aligned} \frac{p(x+1)}{p(x)} &= \frac{(n-x)\theta}{(x+1)(1-\theta)} \quad \text{[Chapter 9, Example 2]} \\ &= \frac{\lambda - x\theta}{(x+1)(1-\theta)} \\ &\to \frac{\lambda}{x+1} \quad \text{as } n \to \infty \text{ and } \theta \to 0 \text{ s.t. } n\theta = \lambda \end{aligned}$$

This is the same ratio as if $X \sim \text{Po}(\lambda)$. Since, in addition, $\sum_{x=0}^{n} p(x) = 1$, then each $p(x)$ must be approximately equal to the corresponding value from the Poisson distribution.

EXERCISES ON 10.1

1. Draw up a table of the first six values of the probability distribution and the cumulative distribution function of the Po(0.5) distribution.
2. An enthusiastic (but not very skilful) crossword-puzzler has examined his attempts at his favourite weekly puzzle. The number of mistakes (including unanswered clues) he makes in a randomly selected puzzle appears to be a Poisson random variable, $X \sim \text{Po}(1)$. The numbers of mistakes made in different puzzles are apparently independent.
   (a) Find the probability that he does not get the next puzzle he tries completely correct.
   (b) Derive the probability distribution of Y, the number of puzzles he must attempt until he first gets one completely correct.
3. A zoologist investigating the marine environment of a beach believes that the distribution of clumps of seaweed satisfies the requirements for a spatial Poisson process. This means that the number of clumps of seaweed in an area of t square metres is a $\text{Po}(10t)$ random variable. She throws a quadrat, a square wooden frame whose sides are each of length $\frac{1}{2}$ metre, onto the beach at a random point and collects all the clumps of seaweed in the frame. Find the probability that she collects at least five clumps of seaweed.
4. The probability that a catastrophic accident occurs in a certain kind of nuclear plant in any given year is claimed to be 'one in a million'. Suppose that 100 similar plants are built world-wide. Find approximately the probability that at least one accident occurs somewhere in the world in the space of ten years.

10.2 The Hypergeometric distribution

Example 4

A bag contains 50 beads, which are identical apart from colour; 40 of the beads are red and the remainder blue. One bead is chosen *at random*, its colour is recorded and it is then *replaced* in the bag and mixed in thoroughly with the others. (This is called **random sampling with replacement**.) The experiment is repeated 15 times in all. If we let the random variable X be the number of times a blue bead is chosen from the bag, then clearly $X \sim \text{Bi}(15, 0.2)$.

Now suppose that, once a bead is drawn from the bag, its colour is recorded but the bead is not returned to the bag. (This is called **random sampling without replacement**.) X is no longer a binomial random variable, for the outcomes of the trials are not independent. For example, the probability of drawing a blue bead on the second draw is either 9/49 or 10/49, depending on whether the first bead drawn is a blue or a red bead. In these circumstances, we say that X is a **hypergeometric** random variable (see also Chapter 8).

This basic 'beads in a bag' experiment can help us to visualize other hypergeometric experiments, such as the following:

(1) At the end of an advertising campaign to promote a new breakfast cereal, the population consists of people who have heard of the new product (like the

blue beads) and people who have not heard of it (like the red beads). If market researchers interview 1000 *different* people (sampling without replacement), then the number of them who have heard of the new product is a hypergeometric random variable.

(2) An auditor checks only a sample of the purchase orders processed by a firm in the course of a year. Since the auditor samples without replacement, then the number of orders in the sample that were handled inappropriately is a hypergeometric random variable.

(3) A student buys a box of ten light bulbs when he moves into a new flat. Unknown to him, two of them are defective. He takes out four bulbs to use immediately; the number of these that are defective is a hypergeometric random variable.

The common features of these experiments can be summarized as follows:

(a) There is a *finite population* of N items ($N = 50$ beads, the population of the UK, the total number of purchase orders, ten light bulbs).

(b) The items are of *two distinct kinds.* M are of type 1 (blue beads, people who have heard of the new product, mishandled orders, defective light bulbs), the remaining $N - M$ are of type 2 (red beads, uninformed people, correctly handled orders, working light bulbs).

(c) A *sample* of n items is drawn from the population *at random and without replacement.*

If the random variable X is the number of items in the sample that are of type 1, then we write $X \sim \text{Hyp}(n, N, M)$. [Note: there is no standard notation for this distribution.] In Example 4, then, $X \sim \text{Hyp}(15, 50, 10)$.

Suppose that $X \sim \text{Hyp}(n, N, M)$. What is its range space? Since there are n items in the sample, $0 \leq X \leq n$. Also, $X \leq M$, since there are only M type 1 items in the whole population. This means that $X \leq \min(n, M)$.

In a similar way, the number of type 2 items in the sample $(n - X)$ must be less than the number of type 2 items in the population $(N - M)$. This means that $X \geq n + M - N$. So, $X \geq \max(0, n + M - N)$.

This means that the range space (R) is the set of all integers between $\max(0, n + M - N)$ and $\min(n, M)$ inclusive. In Example 4, since M $(= 10)$ is less than n $(= 15)$, then $R = \{0, 1, 2, \ldots, 10\}$.

Consider any $x \in R$. There are:

$\binom{N}{n}$ different ways of sampling n items from the population;
$\binom{M}{x}$ different ways of sampling x type 1 items;
$\binom{N-M}{n-x}$ different ways of sampling $n - x$ type 2 items.

Since the sample is drawn at random, all the outcomes are equally likely. So

$$p(x) = P(X = x) = \binom{M}{x}\binom{N-M}{n-x}\bigg/\binom{N}{n} \qquad x \in R$$

This is a valid probability distribution since

(a) $0 \le p(x) \le 1$ for any real value x;
(b)

$$\begin{aligned}\sum_x p(x) &= \frac{1}{\binom{N}{n}} \sum_{x \in R} \binom{M}{x}\binom{N-M}{n-x} \\ &= \frac{1}{\binom{N}{n}} \sum_{x=0}^{n} \binom{M}{x}\binom{N-M}{n-x} \\ &= 1 \quad \text{(Exercise 4 on p. 32)}\end{aligned}$$

Example 5

Car gearboxes are shipped to spares distributors in lots of 50. Before a consignment is sent, an inspector chooses five of the gearboxes at random to be tested. If no more than one of the tested gearboxes is found to be defective, then the consignment is sent as planned. Otherwise, every gearbox is tested and 50 non-defective gearboxes are sent. Find the probability that 100% inspection is required when a lot actually contains three defective gearboxes.

SOLUTION

Let X be the number of defective gearboxes among the five that are tested. Then, $X \sim \text{Hyp}(5, 50, 3)$. So, $X \ge \max(0, 5+3-50) = 0$, and $X \le \min(5, 3) = 3$. In other words, $R = \{0, 1, 2, 3\}$. Also,

$$p(x) = P(X = x) = \binom{3}{x}\binom{47}{5-x} \bigg/ \binom{50}{5} \qquad x = 0, 1, 2, 3$$

We require to find

$$\begin{aligned}P(X > 1) &= \left[\binom{3}{2}\binom{47}{3} + \binom{3}{3}\binom{47}{2}\right] \bigg/ \binom{50}{5} \\ &= [(3 \times 16{,}215) + 1081]/2{,}118{,}760 = 0.0235\end{aligned}$$

This is an example of a quality assurance technique called acceptance lot sampling, which is described in greater detail in Chapter 17.

When a finite population (N) is large, and the sample drawn from it is small in comparison ($n \ll N$), then it makes little difference whether the sample is chosen with or without replacement. The chances of the same item being drawn twice are very small. When the sample is drawn without replacement, the number of type 1 items that are included in the sample is a $\text{Hyp}(n, N, M)$ random variable. When the sample is drawn with replacement, this random variable has a $\text{Bi}(n, M/N)$ distribution. The argument above justifies the following binomial approximation to the Hypergeometric distribution.

Suppose that $X \sim \text{Hyp}(n, N, M)$ when N is large and n is small in comparison (n less than about $0.1N$). Then, $X \approx \text{Bi}(n, M/N)$ in the sense that

$$p(x) \approx \binom{n}{x}\left(\frac{M}{N}\right)^x \left(1 - \frac{M}{N}\right)^{n-x} \qquad x = 0, 1, \ldots, n$$

This result can be proved mathematically in the limit, as $N \to \infty$. Since binomial probabilities are usually easier to calculate than the equivalent hypergeometric ones, experiments such as sample surveys are often treated as though they were binomial experiments.

We have said here that sampling *with replacement* from a *finite population* is a binomial experiment, while sampling *without replacement* from a *finite population* is a hypergeometric experiment. Notice that sampling *without replacement* from an *infinite population* (e.g. all possible future components off a production line) is a binomial experiment.

In Example 5, N (the lot size) is just 50, so the binomial approximation to the Hypergeometric distribution would not give accurate answers. Here is a variation on the same theme, where use of the binomial approximation is justified.

Example 6

Electronic calculators are shipped to a large retailer in lots of 500 items. When the consignment is received, 25 calculators are chosen at random and inspected. The lot is rejected if two or more of the tested calculators are found to be defective. Find the probability that a lot containing 20 defective calculators is rejected by the retailer.

SOLUTION

Here X, the number of defective calculators tested, is a Hyp(25, 500, 20) random variable. N (= 500) is reasonably large and n (= 25) is small in comparison, so we will use the binomial approximation to calculate the required probability, $P(X \geq 2)$.

$X \approx \text{Bi}(25, 20/500)$, that is $X \approx \text{Bi}(25, 0.04)$. So,

$$P(X \geq 2) = 1 - P(X = 0) - P(X = 1) = 1 - (0.96)^{25} - 25(0.04)(0.96)^{24}$$
$$= 0.2642$$

(The correct answer, using the original Hypergeometric distribution, is 0.2637.)

EXERCISES ON 10.2

1. Suppose that $X \sim \text{Hyp}(3, 15, 5)$. Find the range space and the probability distribution of X.
2. Suppose that a lawyer has been guilty of financial irregularities in two out of the 200 client accounts the lawyer controls. An auditor will randomly sample 20 accounts to inspect in detail. Find the probability that the auditor investigates at least one of the irregular accounts.
   (a) Work directly, using the appropriate Hypergeometric distribution.
   (b) Use a binomial approximation.
3. Out of a week's production of 5000 car batteries, 100 are defective. 250 of the batteries produced that week are sold to a spares distributor. Find the (approximate) probability that no more than four of them are defective.
4. A university has 2000 students living in the accommodation it owns; 20 of them fail to pay their rent in full by the end of the session. Assuming that the defaulters are randomly distributed around the university's residences, find

the approximate probability that there are more than two defaulters in a particular residence that houses 300 students.

Application: counting endangered species

It is well known that a number of animal species are under threat from human exploitation. These include whales, hunted for food and oil products, and elephants, hunted for ivory. Various governmental and non-governmental organizations are attempting to protect these species. It is important for them to know how many animals of the threatened species are living in a particular geographical location at a particular time. This information can be compared with similar data from previous years to check whether or not the population size has been stabilized.

It is not easy to count animals directly. Not only do they move about and hide, but they can be difficult to tell apart! Various indirect methods to determine the population size have been developed. Here is a simple example of **the capture–recapture method**.

In a certain African nature reserve, there are an unknown number of elephants (N, say, where N is believed to be of the order of a few hundred). The rangers set out to determine N as follows:

(1) On a particular day, they capture and mark 20 of the elephants they encounter at various sites around the reserve.
(2) They release these elephants and allow them to mingle with the rest of the herd again for one week.
(3) The rangers then capture and examine a second sample of 25 elephants. The random variable X is the number of these elephants that had been marked earlier. It follows that X has a Hypergeometric distribution, $X \sim \text{Hyp}(25, N, 20)$ (see the diagram below).

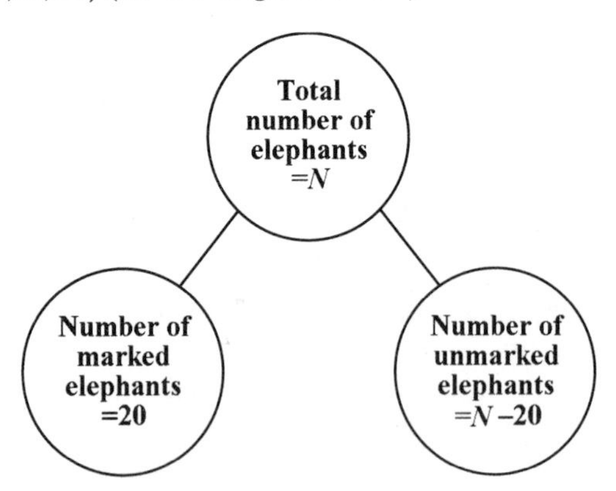

So,

$$P(X = x) = \binom{20}{x}\binom{N-20}{25-x} \bigg/ \binom{N}{25} \qquad x = 0, 1, \ldots, 20$$

Suppose that X is observed to take the value x. N is unknown, and clearly $P(X = x)$ is different for different values of N. A *small* value of x makes *large* values of N seem more plausible, for the larger the herd size the less likely it is to capture the same elephants in both samples. On the other hand, a *large* value of x makes *small* values of N seem more plausible, for the same elephant is more likely to be captured twice when the herd is small.

It can be shown that a good estimate for N satisfies the condition $x/25 = 20/N$. In other words, the proportion of marked elephants in the second sample is the same as the proportion of marked elephants in the whole herd. In practice, N is actually estimated by the largest integer less than or equal to $500/x$. Some examples are given in Table 10.2.

Table 10.2 Estimates of herd size for various possible observed values of x.

Observed x	Estimate of herd size —
0	∞
1	500
2	250
3	166
4	125
5	100

There are known to be inadequacies in this experimental method. For example, it is unlikely that the second sample is really drawn at random, since the marked animals will not have mixed thoroughly with the herd again. Nevertheless, the capture–recapture method has proved enduringly popular with zoologists, who face considerable practical difficulties in counting animals.

Recently, there has been a new interest in using this method (which is capable of considerable refinement) to estimate the numbers of people in outcast social groups such as the homeless mentally ill and street prostitutes (Fisher *et al.*, 1994; Laporte, 1994; McKeganey *et al.*, 1992).

TUTORIAL PROBLEM 1

If there are black taxis in your area, design a capture–recapture experiment to estimate the total number of them. Alternative 'species' could be buses run by a particular operator or fleet cars with a distinctive livery. Give some thought to your experimental method; for example, the number of these vehicles on the road might depend on the time of day and the day of the week. After conducting your experiment, compare your answers with others in your tutorial or lecture group. Can you find out from the fleet operator the actual number of vehicles in the population?

Summary

This chapter has introduced two further families of distributions for discrete random variables. The Poisson distribution is used to model the count of events that occur in continuous time or space. It is a limiting form of the Binomial distribution as n becomes large and θ becomes small, but in such a way that $n\theta$ remains constant at a moderately large value. The Hypergeometric distribution describes the number of items of a specified kind in a random sample that is drawn without replacement from a finite population. The Hypergeometric distribution is replaced by the Binomial distribution when the sampling experiment is conducted with replacement. For this reason, the Binomial distribution can be used to approximate the Hypergeometric distribution when the population is large and the sample drawn from it is relatively small.

FURTHER EXERCISES

1. Vehicles travelling in one particular direction along a local road pass at the average rate of nine vehicles per minute at peak times. A pedestrian who wishes to cross safely to a traffic island in the middle of the road needs an interval of five seconds in which no vehicle passes. Assuming that the passage of cars meets the requirements for a Poisson process, find the probability that no vehicle passes in an interval of five seconds.
2. The process of making cut crystal vases can lead to air bubbles being trapped in the crystal. A particular glass blower believes that the number of air bubbles he traps in a vase is a Po(0.5) random variable. Vases in which no air bubbles are trapped are sold as perfect; vases with one or two air bubbles are sold as 'seconds'; other vases cannot be sold. What proportion of vases of each type does this glass blower produce?
3. A certain historical document is known to have been written by either author A or author B. Until now, it was thought equally likely to have been written by either author. Further evidence of authorship is now being sought in the way that the rare word 'upon' is used in the disputed document. The number of times that a rare word occurs in a fixed length of text can be assumed to be a Poisson random variable.

   In undisputed works of similar length to the disputed text, author A uses 'upon' five times on average, while author B uses it once on average. The word occurs twice in the disputed text. Find the (posterior) probability that it was written by author A.
4. A certain disease affects one person in 10,000. What is the (approximate) probability that two or more people are affected by this disease in a Health Board area which has a population of 100,000?
5. The seeds sold by a certain grower are known to be extremely reliable, only one in a hundred failing to germinate. The seeds are sold in packets of 100. Find (approximately) the probability that at least 97 seeds in a packet germinate.
6. A student believes that the topics for the eight questions in a forthcoming examination will be chosen at random (and, of course, without replacement) from a list of 15 topics covered in the course. The student has time to revise

only nine topics before the exam. What is the probability that at least four of the topics the student revises will be examined?

7. A capture–recapture experiment is carried out to estimate N, the (unknown) number of fish in a pond. First, a random sample of 15 fish are captured, marked and thrown back. After a few days, a second random sample of 25 fish are caught. Write down the probability distribution of X, the number of marked fish in the second sample.
(a) Suppose that X is observed to take the value 3. Write down the probability of this event, p_N, in terms of N.
(b) A sensible estimate of N is N_0, the integer value that maximizes p_N. (This is known as maximum likelihood estimation, a standard statistical technique.) Show that, for all possible N,

$$\frac{p_N}{p_{N-1}} = \frac{N^2 - 40N + 375}{N^2 - 37N}$$

so that $p_N < p_{N-1} \Leftrightarrow 375/3 < N$. Hence find N_0.

11 • Moments

We can find any probability associated with a discrete random variable from either its probability distribution or its cumulative distribution function. Two values are often used to summarize the information contained in these functions. First, there is the expected value, which indicates the typical value of the random variable; it can be interpreted as a long-run average. Secondly, there is the variance, which indicates the spread of values taken by the random variable. These are examples of moments. All the moments of a probability distribution can be found from its moment-generating function.

11.1 The expected value

Example 1

Assume that a liveborn child is equally likely to be a girl or a boy and that the sexes of children from different births are independent (even within the same family). Then X, the number of girls in a family of exactly three children with no multiple births, is a $\text{Bi}(3, \frac{1}{2})$ random variable. Its probability distribution is given below.

x	0	1	2	3
$p(x)$	1/8	3/8	3/8	1/8

Suppose we were to observe X for each of many families, N (say), which had three children. Then the relative frequency definition of probability suggests that about $N/8$ families would have no girls, about $3N/8$ would have one girl, about $3N/8$ would have two girls and about $N/8$ would have three girls. So, the total number of girls in the N families would be about

$$(0 \times N/8) + (1 \times 3N/8) + (2 \times 3N/8) + (3 \times N/8)$$

The average value of X (the average number of girls per family) would be about

$$\begin{aligned}&[(0 \times N/8) + (1 \times 3N/8) + (2 \times 3N/8) + (3 \times N/8)]/N \\ &\qquad = (0 \times 1/8) + (1 \times 3/8) + (2 \times 3/8) + (3 \times 1/8) = 1.5\end{aligned}$$

We call 1.5 the *expected value* of X.

In general, if X is a discrete random variable with range space R and probability distribution $p(x)$, then its **expected value** is defined to be

$$E(X) = \sum_{x \in R} x \cdot p(x)$$

The expected value of X is well defined (i.e. finite) as long as the series $\sum_{x \in R} x \cdot p(x)$ is absolutely convergent (see Hirst (1995)). This will be true for all the random variables discussed in this book.

In Example 1, though it was useful to introduce the concept of the expected value by considering a large number of replicates of a stochastic experiment, N was not needed for calculating $E(X)$. After cancelling all the N, the expected value was found from a particular case of the above formula:

$$[0 \times p(0)] + [1 \times p(1)] + [2 \times p(2)] + [3 \times p(3)]$$

In fact, if we actually recorded the number of girls in many families with three children, the average number of girls would not be exactly 1.5, though it would be very close to that value. The expected value is a theoretical property of a probability distribution, not a property of a sample (however large). It can be interpreted, though, as the average or typical value of the random variable in infinitely many replicates of the underlying experiment. This example also shows that the expected value of X need not be a value in the range space of X; after all, no family can have exactly 1.5 girls!

Example 2

Small particles can be found in the molten glass from which glass bottles are made. If even one of these particles is incorporated into a bottle, then the bottle must be scrapped. Suppose that ten bottles are to be produced from a quantity of molten glass in which two of these particles are randomly distributed. What is the expected number of bottles that will have to be scrapped?

SOLUTION

Each particle is equally likely to be incorporated into each of the ten bottles. This means that, in total, there are $10 \times 10 = 100$ equally likely outcomes of the experiment of distributing the particles among the bottles. Exactly ten of these outcomes result in both particles being incorporated in the same bottle; the remainder result in the two particles being in different bottles.

Let the random variable X be the number of bottles in which there is at least one particle. The probability distribution of X is

x	1	2
$p(x)$	0.1	0.9

Therefore

$$E(X) = (1 \times 0.1) + (2 \times 0.9) = 1.9$$

This suggests that, if we carried out this process a large number of times, then we would have to scrap about 1.9 bottles on average each time.

Example 3

Example 1 concerned a particular binomial random variable. Now suppose, in general, that $X \sim \text{Bi}(n, \theta)$; what is its expected value?

$$
\begin{aligned}
E(X) &= \sum_{x=0}^{n} x \cdot p(x) \\
&= \sum_{x=1}^{n} x \cdot \frac{n!}{x!(n-x)!} \cdot \theta^x \cdot (1-\theta)^{n-x} \\
&\qquad\qquad\qquad\qquad \text{(since the term at } x = 0 \text{ is equal to zero)} \\
&= n\theta \cdot \sum_{x=1}^{n} \frac{(n-1)!}{(x-1)!(n-x)!} \cdot \theta^{x-1} \cdot (1-\theta)^{n-x} \\
&= n\theta \cdot \sum_{x=1}^{n} \binom{n-1}{x-1} \cdot \theta^{x-1} \cdot (1-\theta)^{n-x} \\
&= n\theta \cdot \sum_{y=0}^{n-1} \binom{n-1}{y} \cdot \theta^{y} \cdot (1-\theta)^{(n-1)-y} \quad \text{(where } y = x - 1) \\
&= n\theta \cdot [\theta + (1-\theta)]^{n-1} \quad \text{(Binomial Theorem)} \\
&= n\theta
\end{aligned}
$$

This result makes intuitive sense. Consider the particular case where $n = 100$, $\theta = 1/4$ and $E(X) = 100 \times 1/4 = 25$. If we carried out 100 trials with success probability 1/4 on each, we would expect about 25 successes. In general, if we carried out n trials with success probability θ, then we would expect about $n\theta$ successes.

Let $g(X)$ be any real-valued function of the discrete random variable X. Then, $g(X)$ itself is a random variable, and its expected value is defined to be

$$E(g(X)) = \sum_{x \in R} g(x)p(x)$$

This value is well defined (i.e. finite) as long as the series $\sum_{x \in R} g(x)p(x)$ is absolutely convergent (see Hirst (1995)).

Example 4

Suppose you are playing a board game with an unbiased die. At a particular point in the game, it is your turn to roll the die but you will only be allowed to move if you throw a 6, in which case you will move six squares forward.

Let the random variable X be the score on the die and let $g(x)$ be the number of squares moved as a result of scoring x $(x = 1, 2, \ldots, 6)$. $E(g(X))$ is the 'average' number of squares you will move in this situation.

x	1	2	3	4	5	6
$g(x)$	0	0	0	0	0	6
$p(x)$	$\frac{1}{6}$	$\frac{1}{6}$	$\frac{1}{6}$	$\frac{1}{6}$	$\frac{1}{6}$	$\frac{1}{6}$

From the above table, $E(g(X)) = (0 \times \frac{1}{6}) + (0 \times \frac{1}{6}) + (0 \times \frac{1}{6}) + (0 \times \frac{1}{6}) + (0 \times \frac{1}{6}) + (6 \times \frac{1}{6}) = 1$. 'On average', then, you will move one square forward on your turn.

EXERCISES ON 11.1

1. Suppose that the discrete random variable X has range space $R = \{-1, 0, 1\}$. Calculate the expected value of X assuming each of the following probability distributions in turn.

x	-1	0	1
$p_1(x)$	0.25	0.5	0.25
$p_2(x)$	0.5	0.25	0.25
$p_3(x)$	0.25	0.25	0.5

2. Suppose that, in Example 2, there were three particles in the molten glass instead of two. Show that the expected number of scrapped bottles would be 2.71.
3. Suppose that the random variable X has a $\text{Po}(\theta)$ distribution. Show that $E(X) = \theta$. Notice that, when introducing the Poisson distribution, θ was defined to be the average rate per time interval at which events were occurring, which anticipates this result. [You will need to use a similar trick to the one used to derive the expected value of the binomial random variable in Example 3.]

11.2 The variance

Example 5

Wishing to buy a bag of five oranges at your local cut-price food store, you have a choice between two different brands. Let the random variables X_1 and X_2, respectively, be the numbers of inedible oranges in bags of five of the two different types. As a regular purchaser of oranges, you believe from previous experience that X_1 and X_2 have the following probability distributions:

x	0	1	2	3	4	5
$p(x_1) = P(X_1 = x)$	0.6	0.3	0.1	0	0	0
$p(x_2) = P(X_2 = x)$	0.9	0	0	0	0	0.1

Which oranges would you prefer to buy (assuming that you cannot always tell that an orange is inedible just by looking at it)? The expected numbers of inedible oranges are

$$E(X_1) = (0 \times 0.6) + (1 \times 0.3) + (2 \times 0.1) = 0.5$$
$$E(X_2) = (0 \times 0.9) + (5 \times 0.1) = 0.5$$

Though the expected number of inedible oranges per bag is 0.5 in each case, the two probability distributions are very different. This shows that the expected value on its own cannot summarize a probability distribution adequately. As well as an indication of the typical result of an experiment, we would like to know something about the spread of probable values. The value most usually quoted for this purpose is the variance.

Let X be a discrete random variable with probability distribution $p(x)$. Then, the **variance** of X is defined to be

$$V(X) = E[X-\mu)^2] = \sum_{x\in R}(x-\mu)^2 p(x)$$

where $\mu = E(X)$. The variance is finite as long as the series $\sum_{x\in R}(x-\mu)^2 p(x)$ is (absolutely) convergent. This will be true for all the random variables discussed in this book.

The variance, then, tells us the average squared distance of the random variable X from its expected value. Since $(x-\mu)^2$ is non-negative for every $x \in R$, it follows that the variance is also non-negative, that is $V(X) \geq 0$. Usually, $V(X)$ is positive, but Exercise 1 on p. 95 describes the case in which $V(X) = 0$.

It is often simpler to calculate $V(X)$ using the following formula:

$$\begin{aligned}
V(X) &= E[(X-\mu)^2] \\
&= \sum_{x\in R}(x-\mu)^2 p(x) \\
&= \sum_{x\in R}(x^2 - 2\mu x + \mu^2)p(x) \\
&= \sum_{x\in R} x^2 \cdot p(x) - 2\mu \sum_{x\in R} x \cdot p(x) + \mu^2 \sum_{x\in R} p(x) \\
&= E(X^2) - 2E(X)\cdot E(X) + [E(X)]^2 \cdot 1 \quad (\text{since } \mu = E(X)) \\
&= E(X^2) - [E(X)]^2
\end{aligned}$$

Example 4 (continued)

Using this formula

$$E(X_1^2) = (0^2 \times 0.6) + (1^2 \times 0.3) + (2^2 \times 0.1) = 0.7$$

so

$$V(X_1) = E(X_1^2) - [E(X_1)]^2 = 0.7 - [0.5]^2 = 0.45$$

$$E(X_2^2) = (0^2 \times 0.9) + (5^2 \times 0.1) = 2.5$$

so

$$V(X_2) = E(X_2^2) - [E(X_2)]^2 = 2.5 - [0.5]^2 = 2.25$$

These values reflect the information from the probability distribution about the spreads of the probable values of the two random variables; the probability distribution of X_2 is much more spread out than that of X_1.

The **standard deviation** of X is defined by

$$\text{sd}(X) = \sqrt{V(X)}$$

This is well defined since $V(X) \geq 0$. The standard deviation effectively measures the same property of the probability distribution as the variance. Its advantage is that it is measured in the same units as X itself. For example, if X is a number of oranges, then $V(X)$ has the dimension of oranges^2, while $\mathrm{sd}(X)$ has the dimension of oranges.

Example 3 (continued)

Find $V(X)$ when $X \sim \mathrm{Bi}(n, \theta)$.

$$
\begin{aligned}
E(X^2) &= \sum_{x=0}^{n} x^2 \cdot p(x) \\
&= \sum_{x=0}^{n} [x(x-1) + x] \cdot p(x) \\
&= \sum_{x=2}^{n} x(x-1) \cdot p(x) + \sum_{x=0}^{n} x \cdot p(x) \\
&= n(n-1)\theta^2 \cdot \sum_{x=2}^{n} \binom{n-2}{x-2} \cdot \theta^{x-2} \cdot \theta^{n-x} + E(X) \\
&= n(n-1)\theta^2 \cdot \sum_{y=0}^{n-2} \binom{n-2}{y} \cdot \theta^{y} \cdot (1-\theta)^{(n-2)-y} + n\theta \quad (y = x-2) \\
&= n(n-1)\theta^2 + n\theta \quad \text{(Binomial Theorem)}
\end{aligned}
$$

So, $V(X) = E(X^2) - [E(X)]^2 = n(n-1)\theta^2 + n\theta - (n\theta)^2 = n\theta(1-\theta)$.

An easier method of calculating $V(X)$ in this case is described in the next section.

EXERCISES ON 11.2

1. The 'degenerate' random variable, X, takes the value c with probability 1. Show that $E(X) = c$ and $V(X) = 0$.
2. The discrete random variable, X, has range space $R = \{-1, 1\}$.
   (a) Suppose that $p(1) = p(-1) = \frac{1}{2}$. Find $E(X)$ and $V(X)$.
   (b) Now find $E(X)$ and $V(X)$ in the general case where $p(1) = \theta$ and $p(-1) = 1 - \theta$.
3. Find $V(X)$ for each of the probability distributions in Exercise 1 on p. 93.
4. Suppose that $X \sim \mathrm{Bi}(n, \theta)$. Find $E(X)$, $V(X)$ and $\mathrm{sd}(X)$ when:
   (a) $n = 100, \theta = 0.1$;
   (b) $n = 100, \theta = 0.2$;
   (c) $n = 100, \theta = 0.5$;
   (d) $n = 100, \theta = 0.9$.
5. Show that the variance of a $\mathrm{Po}(\theta)$ random variable is θ. [You will need to use a similar method to that used for the Binomial distribution in Example 3.]

11.3 The moment-generating function

(This section introduces a more advanced topic in probability. It may be omitted on a first reading.)

Perhaps it is already clear that deriving the expected value and variance of special discrete distributions such as the Binomial and Poisson can be difficult. It can often be easier to use a less direct method, which will now be described.

$E(X)$ and $V(X)$ are examples of moments. In general, the **rth moment** ($r = 1, 2, \ldots$) of the random variable X is $E(X^r)$. So, $E(X)$ is the first moment and $E(X^2)$ the second moment of X.

The **rth central moment** of X is $E[(X - \mu)^r]$, $r = 1, 2, \ldots$. So, $V(X)$ is the second central moment of X. The first central moment of a random variable is always 0 (see Exercise 1 on pp. 97–8).

The **moment-generating function (m.g.f.)** of X is defined in terms of the dummy variable u as follows:

$$M_X(u) = E(\mathrm{e}^{Xu}) = \sum_{x \in R} \mathrm{e}^{xu} p(x)$$

$M_X(u)$ is defined for all values of u such that the series $\sum_{x \in R} \mathrm{e}^{xu} p(x)$ is (absolutely) convergent (see Hirst (1995)). $M_X(u)$ is usually (but not always) defined for all real values, u.

Any moment of X can be derived from the m.g.f. as follows:

$$\begin{aligned} M_X(u) &= \sum_x \mathrm{e}^{xu} p(x) \\ &= \sum_x \left(1 + xu + \frac{(xu)^2}{2!} + \frac{(xu)^3}{3!} + \ldots\right) p(x) \end{aligned}$$

(Maclaurin expansion of e^{xu})

$$\begin{aligned} &= \sum_x p(x) + \sum_x xu \cdot p(x) + \sum_x \frac{(xu)^2}{2!} p(x) + \ldots \\ &= 1 + uE(X) + \frac{u^2}{2!} \cdot E(X^2) + \ldots \end{aligned}$$

So,

$$M'_X(u) = \frac{\mathrm{d}}{\mathrm{d}u} M_X(u) = E(X) + uE(X^2) + \frac{u^2}{2!} E(X^3) + \ldots$$

$$M''_X(u) = \frac{\mathrm{d}^2}{\mathrm{d}u^2} M_X(u) = E(X^2) + uE(X^3) + \ldots$$

and, in general, the rth derivative is

$$M_X^{(r)}(u) = \frac{\mathrm{d}^r}{\mathrm{d}u^r} M_X(u) = E(X^r) + uE(X^{r+1}) + \ldots \quad (r = 1, 2, 3, \ldots)$$

Evaluating these derivatives at $u = 0$, it follows that

$$M_X^{(r)}(0) = E(X^r) \quad (r = 1, 2, 3, \ldots)$$

So, another method of calculating the rth moment of X is to find its moment-generating function, differentiate r times and set $u = 0$.

Example 3 (continued)

Suppose that $X \sim \text{Bi}(n, \theta)$. Then, X has m.g.f.

$$\begin{aligned} M_X(u) &= E(\mathrm{e}^{Xu}) \\ &= \sum_{x=0}^{n} \mathrm{e}^{xu} \binom{n}{x} \theta^x (1-\theta)^{n-x} \\ &= \sum_{x=0}^{n} \binom{n}{x} (\theta \mathrm{e}^u)^x (1-\theta)^{n-x} \\ &= (\theta \mathrm{e}^u + 1 - \theta)^n \quad \text{(Binomial Theorem)} \\ M'_X(u) &= n(\theta \mathrm{e}^u + 1 - \theta)^{n-1} \theta \mathrm{e}^u \\ M''_X(u) &= n(n-1)(\theta \mathrm{e}^u + 1 - \theta)^{n-2} \theta \mathrm{e}^u \theta \mathrm{e}^u + n(\theta \mathrm{e}^u + 1 - \theta)^{n-1} \theta \mathrm{e}^u \end{aligned}$$

So,

$$\begin{aligned} E(X) = M'_X(0) &= n(\theta \mathrm{e}^0 + 1 - \theta)^{n-1} \theta \mathrm{e}^0 = n(\theta \cdot 1 + 1 - \theta)^{n-1} \cdot \theta \cdot 1 = n\theta \\ E(X^2) = M''_X(0) &= n(n-1)(\theta \mathrm{e}^0 + 1 - \theta)^{n-2} \theta \mathrm{e}^0 \theta \mathrm{e}^0 + n(\theta \mathrm{e}^0 + 1 - \theta)^{n-1} \theta \mathrm{e}^0 \\ &= n(n-1)(\theta \cdot 1 + 1 - \theta)^{n-2} \cdot \theta \cdot 1 \cdot \theta \cdot 1 \\ &\quad + n(\theta \cdot 1 + 1 - \theta)^{n-1} \cdot \theta \cdot 1 \\ &= n(n-1)\theta^2 + n\theta \end{aligned}$$

$$V(X) = E(X^2) - [E(X)]^2 = n^2\theta^2 - n\theta^2 + n\theta - (n\theta)^2 = -n\theta^2 + n\theta = n\theta(1-\theta)$$

These are the same results as before, but they have been obtained without the need to remember all the tricks required previously.

In general, it is easier to find $E(X)$ and $V(X)$ by direct summation when $P(X = x)$ is specified in a table, but through m.g.f.s when $P(X = x)$ is given by a general formula.

EXERCISES ON 11.3

1. Let X be an arbitrary discrete random variable with range space R and probability distribution $\{p(x), x \in R\}$. Show that the first central moment of X is 0.
2. Suppose that $X \sim \text{Po}(\theta)$. Show that X has m.g.f.

   $$M_X(u) = \exp[\theta(\mathrm{e}^u - 1)]$$

   Hence show that $E(X) = V(X) = \theta$. [You will need to spot the Maclaurin expansion of $\exp(\theta \mathrm{e}^u)$.]
3. Suppose that $X \sim \text{Ge}(\theta)$. Show that X has the following m.g.f.:

   $$M_X(u) = \frac{(1-\theta)\mathrm{e}^u}{1 - \theta \mathrm{e}^u} \qquad u < -\log_e \theta$$

Hence show that

$$E(X) = \frac{1}{(1-\theta)} \quad \text{and} \quad V(X) = \frac{\theta}{(1-\theta)^2}$$

[You will need to know the sum to infinity of a geometric series.]

Summary

Any probability associated with the discrete random variable X can be found using the probability distribution. No single number can adequately summarize all of this information. The expected value, $E(X)$, indicates the typical value that X takes; it can be interpreted as a long-run average. The variance, $V(X)$, is a measure of the spread of X around its expected value. Both $E(X)$ and $V(X)$ can be determined directly from the probability distribution, but it is sometimes easier to evaluate $E(X)$ and $V(X)$ from the moment-generating function instead.

FURTHER EXERCISES

(Exercises below that make use of the moment-generating function are marked with a *.)

1. Find $E(X)$, $V(X)$ and $\text{sd}(X)$ when:
   (a) X is a single random digit (0 to 9 inclusive);
   (b) X is the score obtained on one roll of a fair die;
   (c) X is the sum of the scores obtained on two rolls of a fair die.
2. When $X \sim \text{Hyp}(M, N, n)$, then it can be shown that

$$E(X) = n \cdot \frac{M}{N} \quad \text{and} \quad V(X) = n \cdot \frac{M}{N} \cdot \left(1 - \frac{M}{N}\right) \cdot \frac{N-n}{N-1}$$

   Verify these formulae in the particular cases: (a) $X \sim \text{Hyp}(1, 10, 5)$; (b) $X \sim \text{Hyp}(3, 20, 5)$.
3. A computer supply firm offers an on-site warranty, with the following terms. If your computer develops a hardware fault, then an engineer will call to repair it. If, once the engineer has left, the fault recurs, then an engineer will again call to repair it. If the fault recurs after this, then the firm will call a third time to replace your computer with a similar machine that has been fully tested and is certain to work.

   The firm believes that, at any visit, independently of what has happened previously, one of their engineers has probability 0.9 of effecting a permanent repair on site. Let the random variable X be the total number of calls required to resolve a problem. Write down the probability distribution of X and calculate $E(X)$ and $V(X)$.
4. A child is being taught how to solve a new jigsaw. Since this is a learning exercise, the probability that the child fails to complete the jigsaw falls with each attempt that the child makes. It is believed that the probability that the child fails on the ith attempt is $1/(1+i)$, $i = 1, 2, 3, \ldots$. Let X be the number of successes in the child's first three attempts. Write down the probability distribution of X and calculate $E(X)$ and $V(X)$.

***5.** In Exercise 3 in the Further Exercises section of Chapter 8, the following probability distribution was introduced to model X, the number of children in a completed family:

$$p(0) = \lambda$$

$$p(x) = (1 - \lambda)(1 - \theta)\theta^{x-1} \quad (x = 1, 2, \ldots)$$

Show that

$$M_X(u) = \lambda + \frac{(1 - \lambda)(1 - \theta)e^u}{1 - e^u\theta} \quad (u < -\log_e\theta)$$

$$E(X) = \frac{(1 - \lambda)}{(1 - \theta)} \quad V(X) = \frac{(1 - \lambda)(\lambda + \theta)}{(1 - \theta)^2}$$

12 • Continuous Random Variables

We need continuous random variables to describe experiments which result in a real (rather than an integer) value being recorded, for example a measurement of height, temperature or lifetime. This chapter introduces the key concepts of the (cumulative) distribution function, the probability density function and the moment-generating function.

12.1 The cumulative distribution function

Example 1

Trains on my local underground line are advertised to run every three minutes. I have never bothered to find out when the trains are scheduled to arrive, so I assume that I get to the station at a random point in the timetabled service. X, the length of time (minutes) I must wait until the next train arrives, is a continuous random variable. Its range space is $R = \{x : 0 < x < 3\}$.

The probability that I have to wait between a and b minutes must be proportional to the length of the interval (a, b). For example:

$$P(0 < X < 3) = 1;$$
$$P(0 < X < 1.5) = \tfrac{1}{2};$$
$$P(1 < X < 2) = 1/3;$$

in general, $P(a < X < b) = (b - a)/3$ for $0 < a \le b < 3$.

Suppose that $x \in R$. Let $\delta > 0$ be a value such that $0 < x - \delta \le x + \delta < 3$. Then

$$P(X = x) = \lim_{\delta \to 0} P\{x - \delta < X < x + \delta\} = \lim_{\delta \to 0}(2\delta/3) = 0$$

So, $P(X = x) = 0$ for every $x \in R$, yet some value of x must occur! The solution to this apparent paradox is that an event with probability 0 is not necessarily impossible (though an impossible event necessarily has probability 0); see Chapter 5.

For any continuous random variable, X, the same result holds: the probability that X equals any particular value is 0. Consequently, it is meaningless to try to write down a probability distribution for a continuous random variable. However, it is still meaningful to define the cumulative distribution function:

$$F(x) = P(X \le x) \quad \text{for all real values, } x$$

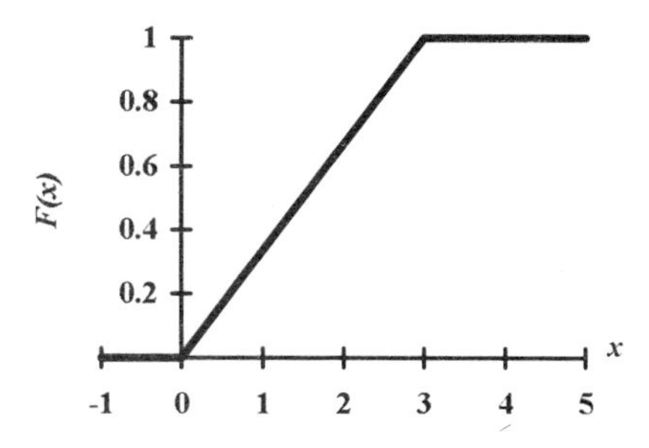

Fig 12.1 The (cumulative) distribution function of the continuous random variable, X, in Example 1.

In Example 1, for instance,

$$F(x) = P(X \leq x) = \begin{cases} 0 & x \leq 0 \\ P(0 < X \leq x) = (x - 0)/3 = x/3 & 0 < x < 3 \\ 1 & x \geq 3 \end{cases}$$

This function is plotted in Fig 12.1. Although the c.d.f. is defined in exactly the same way for a continuous random variable as for a discrete random variable, it is not a step function but is continuous on the range space R. The c.d.f. has the following properties, whether the random variable is discrete or continuous (see also Chapter 8):

(a) $F(-\infty) = 0$; $F(\infty) = 1$;
(b) F is a non-decreasing function: if $a \leq b$, then $F(a) \leq F(b)$.

The c.d.f. is the basic means used to describe the probabilities associated with a continuous random variable. If we know all probabilities of the form $P(X \leq x)$, then we can derive the probability that X lies in any interval:

$$P(a < X \leq b) = P(X \leq b) - P(X \leq a) = F(b) - F(a)$$

Since $P(X = a) = P(X = b) = 0$ when X is a continuous random variable, it follows that

$$\begin{aligned} P(a \leq X \leq b) &= P(a \leq X < b) = P(a < X \leq b) = P(a < X < b) \\ &= F(b) - F(a) \end{aligned}$$

Example 2

Suppose that arrivals at a bank in the early afternoon follow a Poisson process with a mean rate of θ arrivals per minute. This means that the number of customers who arrive in an interval of t minutes is a $\text{Po}(\theta t)$ random variable. Let the random variable X be the amount of time that passes (in minutes) from a given point in time until the next customer arrives. Then X is a continuous random variable with range space $R = \{x : x > 0\}$.

The cumulative distribution function of X is

$$\begin{aligned} F(x) = P(X \le x) &= P(\text{there is at least one arrival in the next } x \text{ minutes}) \\ &= 1 - P(\text{there is no arrival in the next } x \text{ minutes}) \\ &= 1 - P(\text{a Po}(\theta x) \text{ random variable takes the value } 0) \\ &= 1 - e^{-\theta x} \quad x > 0 \end{aligned}$$

The probability that the next customer arrives between five and ten minutes from now is

$$F(10) - F(5) = e^{-5\theta} - e^{-10\theta}$$

EXERCISES ON 12.1

1. Suppose that the continuous random variable X has range space $R = \{x : 0 < x < 1\}$ and cumulative distribution function $F(x) = x^2$ $(0 < x < 1)$. Find $P(X = 0.5)$, $P(X \le 0.5)$, $P(X < 0.5)$, $P(0.2 < X < 0.6)$, $P(X > 0.8)$.
2. Suppose that the continuous random variable X has range space $R = \{x : 0 \le x \le \pi/2\}$ and cumulative distribution function $F(x) = \sin x (0 \le x \le \pi/2)$. Find $P(X < \pi/4)$, $P(\pi/6 < X < \pi/3)$, $P(X > \pi/3)$.
3. In the context of Example 2, let X be the amount of time that passes from a given point in time until the arrival of the second new customer at the bank. Show that X has cumulative distribution function $F(x) = 1 - (1 + \theta x)e^{-\theta x}$ $(x > 0)$.

12.2 The probability density function

Closely related to the cumulative distribution function is another function which also summarizes probabilistic information about a continuous random variable, X. This is the **probability density function (p.d.f.)**, which is defined by

$$f(x) = \frac{d}{dx} F(x)$$

that is,

$$F(x) = \int_{-\infty}^{x} f(t)\, dt$$

Example 1 (continued)

The probability density function of X is

$$f(x) = F'(x) = \begin{cases} 0 & x \le 0 \\ 1/3 & 0 < x < 3 \\ 0 & x \ge 3 \end{cases}$$

This is plotted in Fig 12.2. As in this example, the p.d.f. of a continuous random variable is always 0 outside the range space R; $f(x) = 0$ $(x \notin R)$. Consequently, in future examples, we will only specify $f(x)$ for $x \in R$.

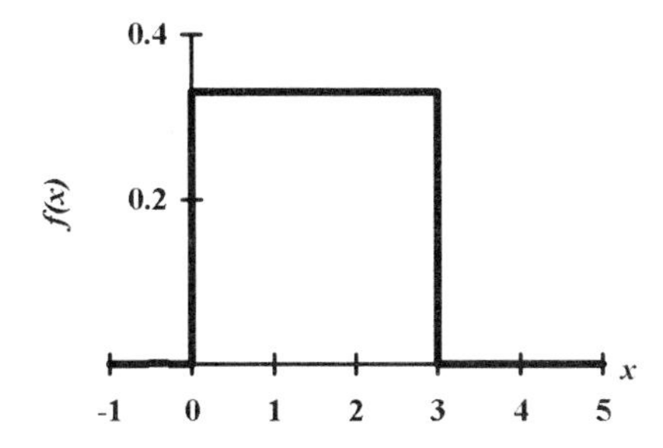

Fig 12.2 The probability density function of the continuous random variable, X, in Example 1.

The probability that the random variable X lies in any interval (a, b) can be found from its p.d.f. as follows:

$$\begin{aligned} P(a < X < b) &= P(X < b) - P(X < a) \\ &= F(b) - F(a) \\ &= \int_{-\infty}^{b} f(x)\,\mathrm{d}x - \int_{-\infty}^{a} f(x)\,\mathrm{d}x \\ &= \int_{a}^{b} f(x)\,\mathrm{d}x \end{aligned}$$

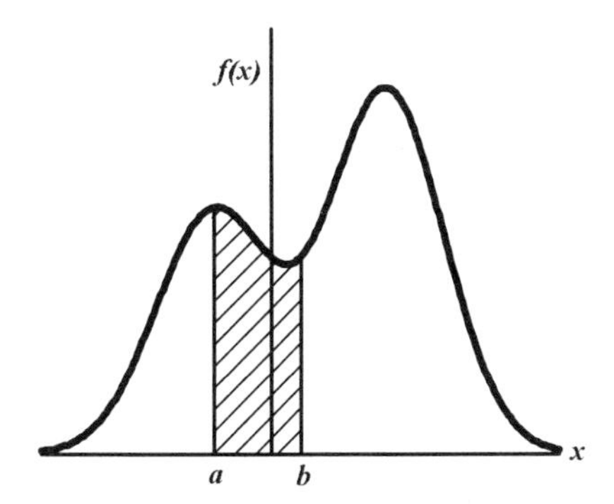

Since $P(X = x) = 0$ for all x, it should be clear that the p.d.f. does not represent probabilities in the same way as the probability distribution of a discrete random variable. There is a connection, though, which can be seen from the following argument.

Let x be any real number and let δ be a small positive value. Then,

$$\begin{aligned} P\left(x - \frac{\delta}{2} < X < x + \frac{\delta}{2}\right) &= \int_{x-\delta/2}^{x+\delta/2} f(t)\,\mathrm{d}t \\ &\approx f(x) \cdot \delta \end{aligned}$$

It must be stressed again that $f(x)$ is NOT the probability that X equals x. However, the p.d.f. is proportional to the probability that X lies in a small interval centred on x. It is for this reason that the p.d.f. is preferred to the c.d.f. as a way of describing the probabilities associated with X. In Example 1, for instance, the p.d.f. is constant in the interval $(0, 3)$, reflecting the intuitive notion that values in this range are 'equally likely'.

X in Example 1 is said to be a **uniform** random variable, written $X \sim \text{Un}(0, 3)$. A uniform random variable can be defined on any finite interval (a, b). We write $X \sim \text{Un}(a, b)$ to mean that X can only take values in the interval (a, b) and has p.d.f.

$$f(x) = \frac{1}{b - a} \qquad a < x < b$$

Example 2 (continued)

If X is the time until the next arrival at the bank, when arrivals occur, according to a Poisson process, at an average rate of θ per minute, then we have already shown that

$$F(x) = 1 - \mathrm{e}^{-\theta x} \qquad x > 0$$

So,

$$f(x) = F'(x) = \theta\mathrm{e}^{-\theta x} \qquad x > 0$$

The probability that the next customer will arrive between one minute and five minutes from a given point in time is

$$\int_1^5 f(x)\,\mathrm{d}x = \int_1^5 \theta\mathrm{e}^{-\theta x}\,\mathrm{d}x = \left[-\mathrm{e}^{-\theta x}\right]_1^5 = \mathrm{e}^{-\theta} - \mathrm{e}^{-5\theta}$$

A continuous random variable X with this p.d.f. is said to be an **exponential** random variable, written $X \sim \text{Ex}(\theta)$. This p.d.f. is sketched in Fig 12.3.

A valid p.d.f. must have the following properties:

(a) $f(x) \geq 0$, for every real value, x. (Since $F(x)$ is a non-decreasing function on R, its first derivative must be non-negative on R.)
(b) $\int_{-\infty}^{\infty} f(x)\,\mathrm{d}x = F(\infty) = 1$.

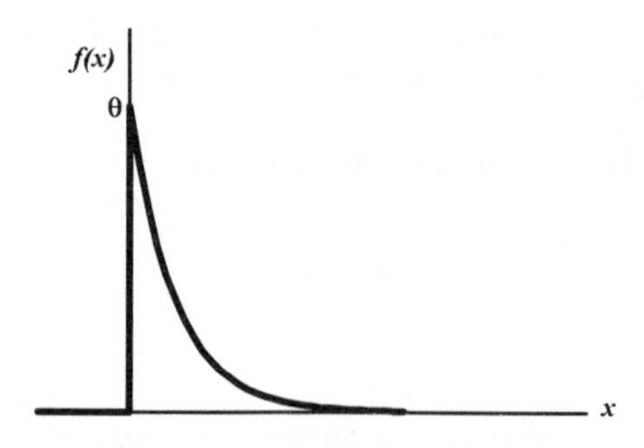

Fig 12.3 The probability density function of an Ex(θ) random variable (Example 2).

EXERCISES ON 12.2

1. Find the probability density functions of the random variables described in Exercises 1, 2 and 3 on p. 102.
2. Either sketch the following functions roughly, or plot them using a graphics calculator or computer. Check that each is a valid probability density function and derive the corresponding cumulative distribution function by integrating.
   (a) $f(x) = 6x(1-x), 0 < x < 1$;
   (b) $f(x) = 3c^3/x^4, x > c$ (where c is a positive real constant);
   (c) $f(x) = 2x\exp(-x^2), x > 0$.
3. Suppose that the continuous random variable Y has range space $R = \{y : 0 < y < 1\}$ and a probability density function of the form

$$f(y) = ky^2(1-y) \quad 0 < y < 1$$

Find the constant, k, that makes this a valid p.d.f. Hence find: (a) $P(Y > 0.6)$; (b) $P(Y < 0.2)$.

12.3 Moments of a continuous random variable

The **expected value** of a continuous random variable is defined in a similar way as that of a discrete random variable (Chapter 11):

$$E(X) = \int_{-\infty}^{\infty} x \cdot f(x)\, dx$$

This value is well defined (i.e. finite) if the integral $\int_{-\infty}^{\infty} x \cdot f(x)\, dx$ is absolutely convergent. As before, $E(X)$ can be thought of as a long-run average; if we carried out the underlying experiment many times, the average recorded value of X would be (about) $E(X)$.

In a similar way, the expected value of a real function, $g(X)$, of the continuous random variable, X, is defined to be

$$E(g(X)) = \int_{-\infty}^{\infty} g(x) \cdot f(x)\, dx$$

The variance of the continuous random variable, X, is defined by

$$V(X) = E[(X-\mu)^2]$$

where $\mu = E(X)$. The variance is finite only if the integral $\int_{-\infty}^{\infty} (x-\mu)^2 f(x)\, dx$ is (absolutely) convergent.

As in the discrete case, it can be shown that (Exercise 1 on p. 107)

$$V(X) = E(X^2) - [E(X)]^2$$

Example 1 (continued)

Here

$$E(X) = \int_{-\infty}^{\infty} x \cdot f(x)\,\mathrm{d}x = \int_0^3 x \cdot \frac{1}{3}\mathrm{d}x = \left[\frac{1}{3}\cdot\frac{x^2}{2}\right]_0^3 = 1.5$$

This is intuitively reasonable; if I wait for my train for a random time between zero and three minutes, then on average I have to wait 1.5 minutes. Also,

$$E(X^2) = \int_{-\infty}^{\infty} x^2 \cdot f(x)\,\mathrm{d}x = \int_0^3 x^2 \cdot \frac{1}{3}\mathrm{d}x = \left[\frac{1}{3}\cdot\frac{x^3}{3}\right]_0^3 = 3$$

so

$$V(X) = E(X^2) - [E(X)]^2 = 3 - [1.5]^2 = 0.75$$

Example 2 (continued)

$$E(X) = \int_{-\infty}^{\infty} x \cdot f(x)\,\mathrm{d}x = \int_0^{\infty} x \cdot \theta \mathrm{e}^{-\theta x}\,\mathrm{d}x$$

Substituting $u = \theta x$, $\mathrm{d}u = \theta\,\mathrm{d}x$, and integrating by parts, this integral equals

$$\begin{aligned}
\int_0^{\infty} \frac{1}{\theta} u\mathrm{e}^{-u}\,\mathrm{d}u &= \frac{1}{\theta}\left\{[u(-\mathrm{e}^{-u})]_0^{\infty} - \int_0^{\infty} 1(-\mathrm{e}^{-u})\,\mathrm{d}u\right\} \\
&= \frac{1}{\theta}\left\{0 + \int_0^{\infty} \mathrm{e}^{-u}\,\mathrm{d}u\right\} \\
&= \frac{1}{\theta}\{[-\mathrm{e}^{-u}]_0^{\infty}\} \\
&= \frac{1}{\theta}
\end{aligned}$$

This means (for example) that if customers arrive at the bank at an average rate of $\theta = 3$ per minute, then the expected time between arrivals is $1/\theta = 1/3$ minute. This is an intuitively appealing result.

Similarly, it can be shown that $E(X^2) = 2/\theta^2$ and so $V(X) = 1/\theta^2$. We will prove these results in another way in Section 12.4.

Example 3

Steel beams eventually break under stress. Electronic components work for a time, then fail. Patients who undergo surgery to remove a tumour might relapse some time later. In all these examples, an important random variable is the **lifetime**, the time from the start of the process (applying the stress, switching the component on, operating) until a failure occurs. It is often assumed that lifetimes follow an $\mathrm{Ex}(\theta)$ distribution, in which case the expected lifetime is $1/\theta$.

Suppose that fuses of a certain kind have an average lifetime of 10,000 hours. Let the random variable T be the lifetime of a fuse. The **reliability** of a fuse, at time t, is the probability that it is still working at time t, which is

$$R(t) = P(T > t) \quad t > 0$$

R is called the **reliability function** in engineering contexts and the **survival function** in medical applications. $R(t) = 1 - F(t)$, where F is the cumulative distribution function of T. Supposing that the lifetime of a fuse is $T \sim \text{Ex}(1/10{,}000)$, then

$$R(t) = e^{-t/10{,}000} \quad t > 0$$

So, the probability that a fuse operates for at least 20,000 hours is

$$R(20{,}000) = e^{-2} = 0.1353$$

and the probability that a fuse operates for at least 50,000 hours is

$$R(50{,}000) = e^{-5} = 0.0067$$

EXERCISES ON 12.3

1. Prove, from the definition of the variance of a continuous random variable, X, that $V(X) = E(X^2) - [E(X)]^2$.
2. Find $E(X)$ and $V(X)$ when $X \sim \text{Un}(a, b)$.
3. Find the expected value and variance of the random variables introduced in Exercise 2 on p. 105. For part (c), you will need to know that

$$\int_0^\infty \sqrt{u} \cdot e^{-u}\, du = \sqrt{\pi}/2$$

4. Let c and d be real constants. Working from the definition of the expected value of a real function of X, show that $E(cX + d) = cE(X) + d$ and that $E[(cX + d)^2] = c^2 E(X^2) + 2cdE(X) + d^2$. Hence show that $V(cX + d) = c^2 V(X)$.
5. Suppose that $X \sim \text{Ex}(\theta)$. Find the probability that X is greater than its expected value.

12.4 The moment-generating function

(This section introduces a more advanced topic in probability. It may be omitted on a first reading.)

The ***r*th moment** of a continuous random variable, X, is defined as

$$E(X^r) = \int_{-\infty}^{\infty} x^r f(x)\, dx$$

As before, the moments of X can be found from the **moment-generating function** (m.g.f.), which for a continuous random variable is defined as follows:

$$M_X(u) = E(e^{Xu}) = \int_{-\infty}^{\infty} e^{xu} f(x)\, dx$$

$M_X(u)$ is defined for all values of u such that the integral $\int_{-\infty}^{\infty} e^{xu} f(x)\, dx$ is (absolutely) convergent.

This formula is identical to the corresponding one for a discrete random variable (Chapter 11), with the p.d.f. replacing the probability distribution and the operation of integration replacing summation. It is still true that $E(X^r)$ can be found by differentiating $M(u)$ r times and setting $u = 0$; in other words,

$$E(X^r) = M_X^{(r)}(0)$$

In particular, $E(X) = M'_X(0)$ and $E(X^2) = M''_X(0)$.

Example 4

When $X \sim \text{Ex}(\theta)$, then

$$M_X(u) = E(\text{e}^{Xu}) = \int_0^\infty \text{e}^{xu}\theta\text{e}^{-\theta x}\,\text{d}x = \theta\int_0^\infty \text{e}^{-(\theta-u)x}\,\text{d}x$$

This integral is infinite unless $-(\theta - u)$ is negative, that is unless $u < \theta$. So, the m.g.f. is only defined when $u < \theta$, but u is arbitrary except that we do want eventually to be able to set $u = 0$. This is all right since $\theta > 0$. So

$$M_X(u) = \theta\int_0^\infty \text{e}^{-(\theta-u)x}\,\text{d}x = \theta\left[-\frac{\text{e}^{-(\theta-u)x}}{\theta-u}\right]_0^\infty = \frac{\theta}{\theta-u} \qquad u < \theta$$

$$M'_X(u) = \theta[-(\theta-u)^{-2}(-1)] = \frac{\theta}{(\theta-u)^2} \qquad u < \theta, \quad \text{so } E(X) = M'_X(0) = \frac{1}{\theta}$$

$$M''_X(u) = \theta[-2(\theta-u)^{-3}(-1)] = \frac{2\theta}{(\theta-u)^3} \qquad u < \theta, \quad \text{so } E(X^2) = M''_X(0) = \frac{2}{\theta^2}$$

$$V(X) = E(X^2) - [E(X)]^2 = \frac{1}{\theta^2}$$

Example 5

Suppose that X has the m.g.f. $M_X(u)$. Define the new random variable Y by $Y = aX + b$, where a and b are real constants. Then Y has m.g.f.

$$M_Y(u) = E(\text{e}^{Yu}) = E(\text{e}^{(aX+b)u}) = E(\text{e}^{bu}\text{e}^{Xau}) = \text{e}^{bu}E(\text{e}^{Xau}) = \text{e}^{bu}M_X(au)$$

The moment-generating function has an important **uniqueness property** that is very useful for proving theoretical results. Suppose that the random variables X and Y have m.g.f.s $M_X(u)$ and $M_Y(u)$ respectively, such that $M_X(u) = M_Y(u)$ for every value u in a range $-\delta < u < \delta$ (for some $\delta > 0$). Then X and Y have the same distribution. This is true whether X and Y are both continuous or both discrete.

Example 4 (continued)

Suppose that $X \sim \text{Ex}(\theta)$. Define the new random variable Y by $Y = kX$, where k is a positive real constant. We can use the uniqueness property of m.g.f.s to prove that $Y \sim \text{Ex}(\theta/k)$.

Notice first that this result is plausible in terms of the expected value and variance of Y. Using Exercise 4 on p. 107:

$$E(Y) = E(kX) = kE(X) = k/\theta \quad \text{and} \quad V(Y) = V(kX) = k^2V(X) = k^2/\theta^2$$

which are respectively the expected value and variance of an $\text{Ex}(\theta/k)$ random variable.

To prove that $Y \sim \text{Ex}(\theta/k)$, we will first find the m.g.f. of Y. We cannot do this directly, since we do not know the p.d.f. of Y. (If we did, there would be nothing left to prove!) Instead, as in Example 5, we find the m.g.f. of Y indirectly from that of X:

$$\begin{aligned} M_Y(u) &= E(\mathrm{e}^{Yu}) \\ &= E(\mathrm{e}^{Xku}) \\ &= M_X(ku) \\ &= \frac{\theta}{\theta - ku} \qquad ku < \theta \\ &= \frac{\theta/k}{\theta/k - u} \qquad u < \theta/k \end{aligned}$$

This is the m.g.f. of an $\text{Ex}(\theta/k)$ random variable, from the previous part of this example. By the uniqueness property of m.g.f.s, then, $Y \sim \text{Ex}(\theta/k)$.

A simpler proof of this result is given in Chapter 14. The uniqueness property of m.g.f.s is useful in another context too (see Chapter 16).

EXERCISES ON 12.4

1. Exercise 3 on p. 102 and Exercise 1 on p. 105 introduced a random variable with probability density function $f(x) = \theta^2 x\mathrm{e}^{-\theta x}$ $(x > 0)$. Using integration by parts, show that

   $$M_X(u) = \left(\frac{\theta}{\theta - u}\right)^2 \qquad u < \theta$$

   Hence find $E(X)$ and $V(X)$.
2. Suppose that $X \sim \text{Un}(a, b)$. Find the moment-generating function of X. Now let $Y = kX$, for some real constant k. Use the moment-generating function to prove that Y is also a uniform random variable.

Summary

This chapter has begun to look more closely at continuous random variables. For a continuous random variable, X, $P(X = x) = 0$ for any real value, x. So, it is pointless to try to define a probability distribution for a continuous random variable. The cumulative distribution function, though, is still well defined. The probability density function can be found from the c.d.f. by differentiating it once. The p.d.f. is the basic tool used to investigate the properties of a continuous random variable; the expected value, variance and other moments are defined in terms of the p.d.f. Again, the moment-generating function has been used as one way to determine the moments of a continuous random variable. The uniqueness property of m.g.f.s has been introduced and exploited for some theoretical purposes.

FURTHER EXERCISES

(Exercises below that make use of the moment-generating function are marked with a *.)

1. Suppose that $X \sim \text{Un}(-0.5, 0.5)$. Find (a) $P(X = 0.5)$; (b) $P(X < 0.5)$; (c) $P(X \leq 0.5)$; (d) $P(X > -0.5)$.
2. Suppose that $X \sim \text{Ex}(3)$. Find (a) $P(X > 1/3)$; (b) $P(1/3 < X < 3)$; (c) $P(X < 3)$.
3. In a recent survey of acute admissions to a geriatric unit in Glasgow, the average time that a patient spent in hospital was recorded as 24 days. If the random variable X is the time spent in hospital by an old person, then it is plausible from the data that $X \sim \text{Ex}(1/24)$. In a follow-up study, it is intended to make measurements on old people after they have been in hospital for two, four and six weeks. What proportion of patients will still be in hospital at each of these times?
4. Let T be the amount of time (years) that elapses between the installation of a certain type of computer hard disk and its failure. T is an $\text{Ex}(1/\lambda)$ random variable.
   (a) Show that, as λ increases, $P(T > t)$ increases for all $t > 0$.
   (b) The manufacturer supplies these hard disks with a warranty for one year. What is the smallest value of λ that a manufacturer must achieve in order to ensure that at most 5% of hard disks are replaced under warranty?
5. Suppose that T, the time to failure of a certain type of electronic component, has the following probability density function:

$$f(t) = \theta\beta t^{\beta-1} \exp(-\theta t^{\beta}) \quad t > 0$$

   where θ and β are positive constants. Prove that T has reliability function, $R(t) = \exp(-\theta t^{\beta})$ for $t > 0$. [T is said to have a Weibull distribution. Notice that the $\text{Ex}(\theta)$ distribution is the special case of the Weibull distribution for which $\beta = 1$.]

*6. Suppose that the random variable X has p.d.f.

$$f(x) = \theta \mathrm{e}^{-\theta(x-c)} \quad x > c$$

   where c is a non-negative real constant. Find the moment-generating function of X, and hence find $E(X)$ and $V(X)$. Compare your answers with those for an $\text{Ex}(\theta)$ distribution.

7. As an extension of Exercise 3 on p. 102, suppose that X is the amount of time that passes until the kth new customer arrives in the bank. Then X has a **Gamma** distribution, written $X \sim \text{Ga}(k, \theta)$.
   (a) Show that

$$F(x) = \sum_{r=k}^{\infty} \frac{\mathrm{e}^{-\theta x}(\theta x)^r}{r!} \quad x > 0$$

   (b) Differentiating term by term, show that

$$f(x) = \frac{\theta^k x^{k-1} \mathrm{e}^{-\theta x}}{(k-1)!} \quad x > 0$$

*(c) Hence prove that

$$M_X(u) = \left(\frac{\theta}{\theta - u}\right)^k \qquad u < \theta$$

[Note that $\int_0^\infty z^{k-1} e^{-z} \, dz = (k-1)!$] Now show that $E(X) = k/\theta$ and $V(X) = k/\theta^2$.

*(d) Show that, if $Y = aX$, for some positive constant a, then $Y \sim \text{Ga}(k, \theta/a)$.

13 • The Normal Distribution

This chapter introduces the most important distribution in probability, the Normal (or Gaussian) distribution. Very many continuous random variables have been found to follow Normal distributions. The Normal distribution can also be used to find approximations to probabilities associated with other distributions, such as the Binomial and the Poisson, as we shall see in Chapter 17.

13.1 The probability density function

Suppose that the random variable, X, can take any real value and that X has p.d.f.

$$f(x) = \frac{1}{\sqrt{2\pi\sigma^2}} \exp\left[-\frac{1}{2}\left(\frac{x-\mu}{\sigma}\right)^2\right] \qquad -\infty < x < \infty$$

(see Fig 13.1). Then X is said to have a **Normal** distribution, with parameters μ and σ^2, written $X \sim \mathrm{N}(\mu, \sigma^2)$. We shall show later that μ is the expected value of X and that σ^2 is its variance.

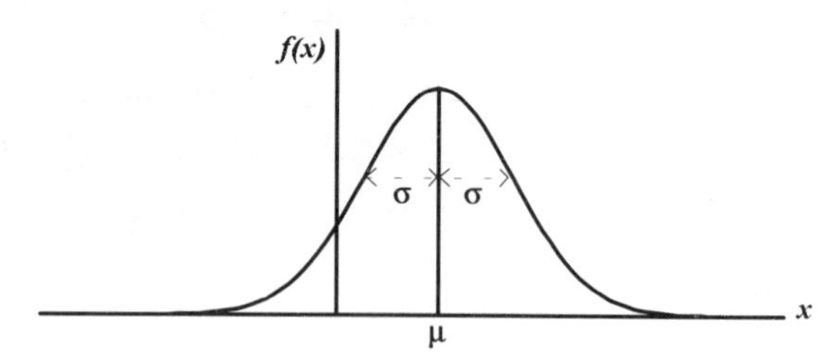

Fig 13.1 The probability density function of the $\mathrm{N}(\mu, \sigma^2)$ distribution.

Many different kinds of continuous random variables have been found to follow Normal distributions, for example:

(1) some anthropometric measurements, such as the height of a fully grown man or woman from a particular racial group;
(2) scientific measurements that are subject to experimental error, for example the measured distance from the Earth to a given star;
(3) dimensions of manufactured objects, for example the weight of a (nominal) $\frac{1}{4}$ lb hamburger.

The two parameters of the Normal distribution allow for a very wide range of probability density functions. The expected value, μ, can be any real number. The standard deviation, σ, can be any positive real number. The larger is σ, the flatter is the probability density function.

As before, we want to be able to calculate probabilities of the form $P(a < X < b)$, which means evaluating integrals of the form

$$\frac{1}{\sqrt{2\pi\sigma^2}} \int_a^b \exp\left[-\frac{1}{2}\left(\frac{x-\mu}{\sigma}\right)^2\right] \mathrm{d}x$$

This cannot be done algebraically, or in closed form, but only using numerical methods of integration.

Tables of the cumulative distribution function of one very important Normal distribution are widely available. This is the N(0, 1) distribution, that is the Normal distribution with expected value 0 and variance 1, which is known as the **Standard Normal** distribution. Appendix 2 at the end of this book contains one such table. This function is so important in practice that we reserve a special notation for it, the capital Greek letter phi (Φ). If $Z \sim \mathrm{N}(0, 1)$, then

$$\Phi(z) = P(Z \leq z) = \frac{1}{\sqrt{2\pi}} \int_{-\infty}^{z} \exp(-x^2/2)\, \mathrm{d}x$$

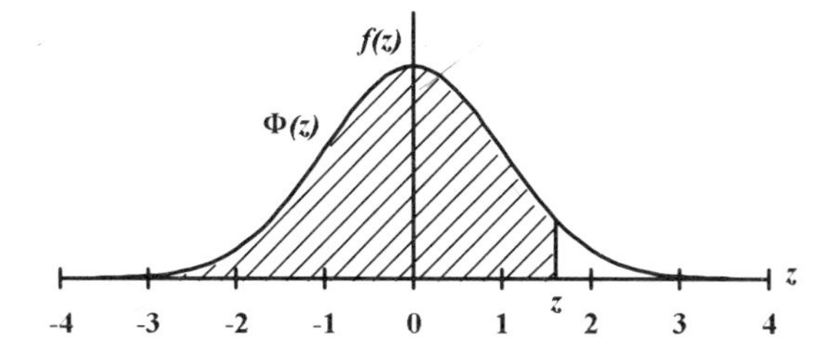

Example 1

Suppose that Z has the Standard Normal distribution, $Z \sim \mathrm{N}(0, 1)$. Then, using Appendix 2 we can easily find probabilities like the following:

$$P(Z \leq 0) = \Phi(0) = 0.5000;$$
$$P(Z \leq 1.96) = \Phi(1.96) = 0.9750;$$
$$P(Z > 1) = 1 - P(Z \leq 1) = 1 - \Phi(1) = 1 - 0.8413 = 0.1587;$$
$$P(1 < Z < 2) = P(Z \leq 2) - P(Z \leq 1) = \Phi(2) - \Phi(1)$$
$$= 0.9772 - 0.8413 = 0.1359$$

Appendix 2 does not tabulate $\Phi(z)$ for $z < 0$. When $Z \sim \mathrm{N}(0, 1)$, then $f(z)$ is symmetric: $f(-z) = f(z)$. It is easily seen, then, that $\Phi(-z) = P(Z \leq -z) = P(Z > z) = 1 - \Phi(z)$, see Fig 13.2.

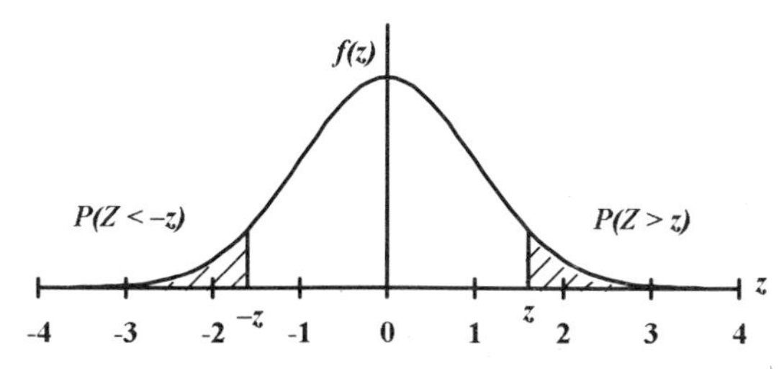

Fig 13.2 The probability density function of the N(0, 1) distribution.

Example 1 (continued)

This result allows us to use Appendix 2 to find probabilities like the following:

$$P(Z < -0.5) = \Phi(-0.5) = 1 - \Phi(0.5) = 1 - 0.6915 = 0.3085;$$
$$P(-0.5 < Z < 0.5) = \Phi(0.5) - \Phi(-0.5) = \Phi(0.5) - [1 - \Phi(0.5)]$$
$$= 2 \cdot \Phi(0.5) - 1 = 0.3830;$$
$$P(-1 < Z < 2) = \Phi(2) - \Phi(-1) = \Phi(2) - [1 - \Phi(1)]$$
$$= 0.9772 - [1 - 0.8413] = 0.8185.$$

This is all very well, as far as it goes, but we still need to work out similar probabilities for other Normal distributions. The following result, which we shall prove in the next section, allows us to do this.

Result 13.1

If $X \sim N(\mu, \sigma^2)$, for any real value, μ, and any positive value, σ, then

$$Z = \frac{X - \mu}{\sigma} \sim N(0, 1)$$

Though we shall not prove this result until later, we can see immediately that the result is correct in terms of the expected value and variance of the two random variables, for (using the results of Exercise 4 on p. 107)

$$E(Z) = E\left(\frac{X-\mu}{\sigma}\right) = E\left(\frac{1}{\sigma}X - \frac{\mu}{\sigma}\right) = \frac{1}{\sigma}E(X) - \frac{\mu}{\sigma} = \frac{\mu}{\sigma} - \frac{\mu}{\sigma} = 0$$

$$V(Z) = V\left(\frac{X-\mu}{\sigma}\right) = V\left(\frac{1}{\sigma}X - \frac{\mu}{\sigma}\right) = \frac{1}{\sigma^2}V(X) = \frac{\sigma^2}{\sigma^2} = 1$$

Example 2

Suppose that $X \sim N(5, 4)$. Find (a) $P(X < 9)$, (b) $P(X > 0)$, (c) $P(-1 < X < 7)$.

SOLUTION

We need to use the result that $Z = (X - 5)/\sqrt{4} = (X - 5)/2 \sim N(0, 1)$.

(a) $P(X < 9) = P[(X - 5)/\sqrt{4} < (9 - 5)/\sqrt{4}] = P(Z < 2) = \Phi(2) = 0.9772.$

(b) $P(X > 0) = P[(X - 5)/\sqrt{4} > (0 - 5)/\sqrt{4}] = P(Z > -2.5) = 1 - \Phi(-2.5)$
$= \Phi(2.5) = 0.9938.$

(c) $P(-1 < X < 7) = P[(-1 - 5)/\sqrt{4} < (X - 5)/\sqrt{4} < (7 - 5)/\sqrt{4}]$
$= P(-3 < Z < 1) = \Phi(1) - \Phi(-3) = \Phi(1) - [1 - \Phi(3)]$
$= 0.8413 - [1 - (0.9987)] = 0.8400.$

Example 3

(a) The actual diameter (in millimetres) of a rivet with nominal diameter 10 mm is an $N(10, 0.01)$ random variable. To be usable, a rivet must have a diameter in the range 9.8 to 10.2 mm. What proportion of rivets are usable?

(b) Suppose now that the diameter of a rivet is an $N(10, \sigma^2)$ random variable. σ, which indicates the accuracy of the manufacturing process, can be modified to some extent. To what value must σ be reduced to ensure that 99% of all rivets are usable?

SOLUTION

(a) Let X be the diameter of a rivet. Then, $X \sim N(10, 0.01)$. So:

$$\begin{aligned} P(9.8 < X < 10.2) &= P\left(\frac{9.8 - 10.0}{0.1} < \frac{X - 10.0}{0.1} < \frac{10.2 - 10.0}{0.1}\right) \\ &= P(-2 < Z < 2) \\ &= 2\Phi(2) - 1 \\ &= 0.9544 \end{aligned}$$

In other words, 95.44% of the rivets are usable.

(b) Now suppose that $X \sim N(10, \sigma^2)$. We must find σ such that $P(9.8 < X < 10.2) = 0.99$.

$$\begin{aligned} P(9.8 < X < 10.2) &= P\left(\frac{9.8 - 10.0}{\sigma} < \frac{X - 10.0}{\sigma} < \frac{10.2 - 10.0}{\sigma}\right) \\ &= P(-0.2/\sigma < Z < 0.2/\sigma) \\ &= 2\Phi(0.2/\sigma) - 1 \end{aligned}$$

We require

$$2\Phi(0.2/\sigma) - 1 = 0.99$$

that is,

$$\begin{aligned} &\Phi(0.2/\sigma) = 0.995 = \Phi(2.57) \quad \text{(from Appendix 2, } \Phi(2.57) = 0.995) \\ &0.2/\sigma = 2.57 \\ &\sigma = 0.0778 \end{aligned}$$

EXERCISES ON 13.1

1. Suppose that $Z \sim N(0.1)$. Find (a) $P(Z < 1.65)$, (b) $P(Z > -1.65)$, (c) $P(-1.65 < Z < 1.65)$, (d) $P(Z < 1.28)$, (e) $P(1.28 < Z < 1.65)$. Also, find the constant $c > 0$ such that $P(-c \leq Z \leq c) = 0.99$.
2. Suppose that $X \sim N(20, 16)$. Find (a) $P(X < 8)$, (b) $P(14 < X < 26)$, (c) $P(X > 30)$.
3. Suppose that $X \sim N(-4, 4)$. Find (a) $P(X < 0)$, (b) $P(-6 < X < -2)$, (c) $P(X > -0.5)$. Also, find the constant c such that $P(X > c) = 0.95$.
4. A consultant in a local hospital can choose between two routes to drive to the hospital. The consultant believes that the travelling time (in minutes) is an $N(40, 25)$ random variable over the first route and an $N(45, 4)$ random variable over the second route. Which is the better route if, in an emergency, the consultant wants to get to the hospital (a) within 45 minutes, (b) within 50 minutes?

13.2 The moment-generating function

(The previous section used some theoretical results that we will now derive formally. This section may be omitted on a first reading.)

In the last section we claimed that, if $X \sim \mathrm{N}(\mu, \sigma^2)$, then $E(X) = \mu$ and $V(X) = \sigma^2$. The integration required to prove this directly is a bit awkward, so we shall use the moment-generating function instead. Suppose that $X \sim \mathrm{N}(\mu, \sigma^2)$. Then,

$$\begin{aligned} M_X(u) &= E(\mathrm{e}^{Xu}) \\ &= \int_{-\infty}^{\infty} \mathrm{e}^{xu} \frac{1}{\sqrt{2\pi\sigma^2}} \exp\left[-\frac{1}{2}\left(\frac{x-\mu}{\sigma}\right)^2\right] \mathrm{d}x \end{aligned}$$

The exponent (power of e) in this integral is

$$\begin{aligned} xu - \frac{1}{2}\left(\frac{x-\mu}{\sigma}\right)^2 &= -[(x-\mu)^2 - 2xu\sigma^2]/2\sigma^2 \\ &= -\{[x - (\mu + u\sigma^2)]^2 - 2\mu u\sigma^2 - u^2\sigma^4\}/2\sigma^2 \\ &\qquad \text{(after completing the square)} \\ &= -\{[x - (\mu + u\sigma^2)]^2\}/2\sigma^2 + \mu u + \frac{1}{2}u^2\sigma^2 \end{aligned}$$

Therefore

$$\begin{aligned} M_X(u) &= \exp\left(u\mu + \frac{1}{2}u^2\sigma^2\right) \int_{-\infty}^{\infty} \frac{1}{\sqrt{2\pi\sigma^2}} \exp\left[-\frac{1}{2}\left(\frac{x - (\mu + u\sigma^2)}{\sigma}\right)^2\right] \mathrm{d}x \\ &= \exp\left(u\mu + \frac{1}{2}u^2\sigma^2\right) \end{aligned}$$

the last step following since the integrand is the p.d.f. of an $\mathrm{N}(\mu + u\sigma^2, \sigma^2)$ random variable.

$$M'_X(u) = \exp(u\mu + \tfrac{1}{2}u^2\sigma^2) \cdot (\mu + u\sigma^2)$$

so

$$E(X) = M'_X(0) = \mu$$

$$M''_X(u) = \exp(u\mu + \tfrac{1}{2}u^2\sigma^2) \cdot (\mu + u\sigma^2)^2 + \exp(u\mu + \tfrac{1}{2}u^2\sigma^2) \cdot \sigma^2$$

(product rule)

so

$$E(X^2) = M''_X(0) = \mu^2 + \sigma^2$$

and

$$V(X) = E(X^2) - [E(X)]^2 = \sigma^2$$

Suppose that $X \sim \mathrm{N}(\mu, \sigma^2)$. Define the random variable $Y = aX + b$, for some real constants a and b. Using the m.g.f. we can show that Y also has a Normal distribution. (We shall prove this result in another way in Chapter 14.) The m.g.f. of Y is

$$M_Y(u) = E(\mathrm{e}^{Yu}) = \mathrm{e}^{bu} M_X(au) \quad \text{(Chapter 12, Example 5)}$$

Since

$$M_X(au) = \exp[(au)\mu + \tfrac{1}{2}a^2u^2\sigma^2]$$

then

$$M_Y(u) = \exp[u(a\mu + b) + \tfrac{1}{2}u^2(a^2\sigma^2)]$$

This is the m.g.f. of an $N(a\mu + b, a^2\sigma^2)$ random variable. By the uniqueness property of the moment-generating function, it follows that $Y \sim N(a\mu + b, a^2\sigma^2)$.

This is a general result. If we let $a = 1/\sigma$ and $b = -(\mu/\sigma)$, then $Z = aX + b = (X - \mu)/\sigma$ and $Z \sim N(0, 1)$, as was claimed in the previous section.

Application: medical diagnosis

The medical diagnosis of a disorder is often guided by the result of a biochemical test based, for example, on the concentration of some blood chemical. Some mistakes will always be made, but the number of them is minimized in a good test.

A subject can be said to be positive (+ve) if the subject has the disorder in question and negative (−ve) otherwise. The subject tests +ve if the test suggests the disorder is present and −ve otherwise. The possible relationships between a person's true status and the outcome of the test are summarized in the following table of *conditional* probabilities.

		Test result	
		+ve	−ve
True status	+ve	θ_+	$1 - \theta_+$
	−ve	$1 - \theta_-$	θ_-

Here, θ_+ is the *conditional* probability that a test result is +ve *given* that the person has the disorder. This is called the **sensitivity** of the test. θ_-, the *conditional* probability that the test result is −ve *given* that the person tested does not have the disorder, is called the **specificity** of the test. The probabilities $1 - \theta_+$ and $1 - \theta_-$ are the error rates; a good test has high sensitivity and specificity.

Crossley *et al.* (1991) describe a pre-natal test for Down's syndrome that is based on the maternal blood serum level of human chorionic gonadotrophin (hCG). The random variable, X, used in their research is the common logarithm of maternal hCG, expressed in multiples of the median in normal controls. Based on blood samples from 459 mothers, these authors suggest the following model for X (see Fig 13.3):

$X \sim N(0, 0.25^2)$ for mothers whose children do not have Down's syndrome;
$X \sim N(0.34, 0.25^2)$ for mothers whose children have Down's syndrome.

In general, the maternal serum levels are higher in mothers of Down's syndrome children than in mothers of unaffected children, but there is considerable overlap between the two groups. A sensible form of **decision rule** is: the child has Down's syndrome if $X > c$; the child does not have Down's syndrome, if $X \leq c$; where c, the **critical value** of the test, is to be determined.

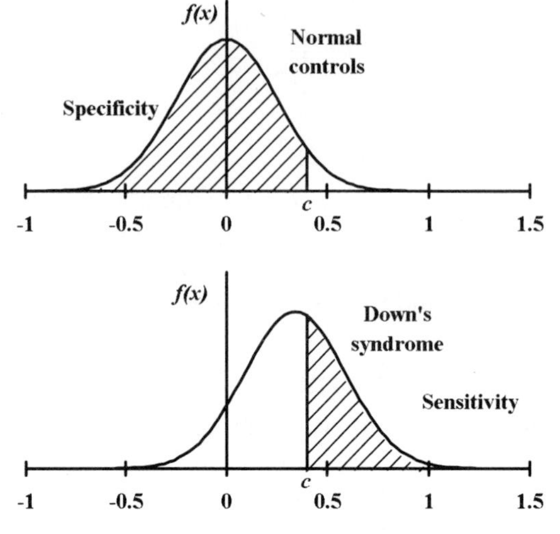

Fig 13.3 The population distributions of maternal serum hCG.

As indicated in Fig 13.3, any choice of c will produce errors. For,

$$\begin{aligned}
\theta_- &= P(\text{test result is } -\text{ve} \mid \text{child does not have Down's syndrome}) \\
&= P(X \le c \mid X \sim \mathrm{N}(0, 0.25^2)) \\
&= P\left(\frac{X-0}{0.25} \le \frac{c-0}{0.25} \;\middle|\; \frac{X-0}{0.25} \sim \mathrm{N}(0,1)\right) \\
&= \Phi(4c) \\
\theta_+ &= P(\text{test result is } +\text{ve} \mid \text{child has Down's syndrome}) \\
&= P(X > c \mid X \sim \mathrm{N}(0.34, 0.25^2)) \\
&= P\left(\frac{X-0.34}{0.25} > \frac{c-0.34}{0.25} \;\middle|\; \frac{X-0.34}{0.25} \sim \mathrm{N}(0,1)\right) \\
&= 1 - \Phi\left(\frac{c-0.34}{0.25}\right) \\
&= 1 - \Phi(4c - 1.36)
\end{aligned}$$

For example, for $c = 0$, the sensitivity of the test would be $1 - \Phi(4c - 1.36) = 1 - \Phi(-1.36) = \Phi(1.36) = 0.9131$. In other words, an impressive 91% of all Down's children would test +ve. However, when $c = 0$, the specificity of the test is $\Phi(4c) = \Phi(0) = 0.5$. So, only 50% of children who do not have Down's syndrome would test −ve. Another way of saying this is that there would be a 50% false +ve rate. This is unacceptably high.

There is a trade-off between specificity and sensitivity. This can be seen from Fig 13.3. Moving c to the left increases the sensitivity (the proportion of Down's syndrome children detected) but reduces the specificity (so more children without Down's syndrome are incorrectly classified).

A **receiver operating characteristic (ROC) plot** graphs sensitivity (θ_+) against 1−specificity ($1 - \theta_-$) for various values of c; a smooth curve is then drawn through these points. Figure 13.4 displays the ROC plot for the current example. As c is

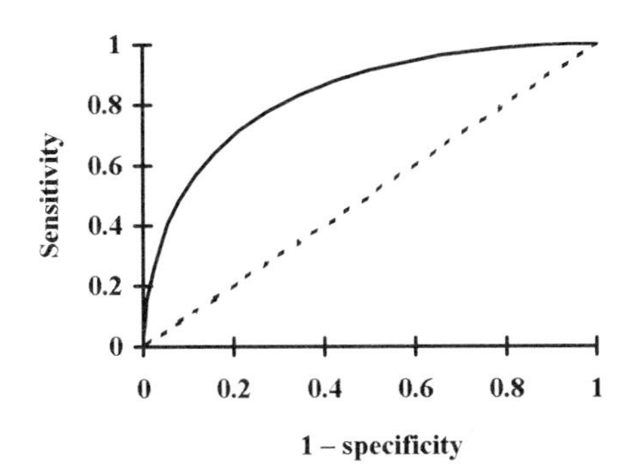

Fig 13.4 ROC plot for the use of maternal serum hCG to detect Down's syndrome. After Crossley *et al.* (1991).

increased from $-\infty$ to $+\infty$, the curve is traced down from the top right-hand corner (sensitivity = 1, specificity = 0) to the bottom left-hand corner (sensitivity = 0, specificity = 1).

The actual choice of c would depend on the relative importance (or cost) of the two types of error. Here, a sensitivity of 0.6 would correspond to a false +ve rate of about 0.14 (i.e. a specificity of about 0.86), which might be an acceptable compromise.

For much more detail about this example, see Crossley *et al.* (1991). For further examples of the use of ROC curves, see Zweig and Campbell (1993).

Summary

This chapter has described the Normal distribution, which is extremely widely used to describe continuous random variables. The Standard Normal distribution, which has expected value 0 and variance 1, is a particularly important case. Any normally distributed random variable may be transformed to the Standard Normal by subtracting its expected value and dividing by its standard deviation. This means that a table of the cumulative distribution function of the Standard Normal distribution (Appendix 2) can be used to calculate probabilities associated with an arbitrary normal random variable. The moment-generating function of the Normal distribution has been derived and used to obtain a number of useful theoretical results. As an example of the way the Normal distribution is used, an application to medical diagnosis has been described.

FURTHER EXERCISES

1. Suppose that $X \sim \mathrm{N}(\mu, \sigma^2)$. Find (a) $P(\mu - \sigma < X < \mu + \sigma)$, (b) $P(\mu - 2\sigma < X < \mu + 2\sigma)$, (c) $P(\mu - 3\sigma < X < \mu + 3\sigma)$.
2. (a) A bagging machine produces a box of breakfast cereal whose weight (in grams) is an $\mathrm{N}(\mu, 100)$ random variable, where μ can be adjusted by the operator. The producer wishes to ensure that 99% of boxes contain at least 500 g of cereal. To what value of μ must the machine be set?

(b) Suppose instead that the standard deviation of the weight of a box of cereal is proportional to the expected weight, μ. Specifically, suppose that the weight of a box is an $N(\mu, (0.02\mu)^2)$ random variable. Find μ so that 99% of boxes contain at least 500 g of cereal.

3. A communication system transmits binary data (sequences of zeros and ones). When a zero is to be communicated, the system sends a voltage signal at level $+1$; when a one is to be communicated, it sends a voltage signal at level -1. Owing to noise on the transmission channel, the level of the received signal is a random variable, X, where $X \sim N(1, \sigma^2)$ when a zero is sent, and $X \sim N(-1, \sigma^2)$ when a one is sent.

The receiver decides that a zero was sent if and only if $X > c$.

(a) Write down a general expression for the probability that a zero is sent but a one is received and for the probability that a one is sent but a zero is received.

(b) Assuming that zeros and ones are sent equally often, use the Law of Total Probability to derive the probability that an error is made in transmitting a digit.

(c) Evaluate these probabilities when $c = 0$, $c = -1$ and $c = 1$.

14 • Functions of a Random Variable

In earlier chapters, we used the moment-generating function to find the distributions of particular functions of some specific random variables. For example, we showed that, if X is normally distributed, then $Y = aX + b$ is also normally distributed. This chapter presents some general methods for determining the distribution of a function of a random variable.

14.1 Discrete random variables

Example 1

To determine whether or not they each have a certain disease, samples of blood are taken from each of ten individuals. Rather than testing each sample separately at the start, subsamples of all ten blood samples are pooled together and the composite sample is tested. If this sample tests negative, then none of the ten individuals has the disease. If it tests positive, then at least one of the ten has the disease and tests must be carried out separately on each specimen to determine precisely who. If 10% of all individuals tested actually have this disease, independently of all other individuals tested, then what is the expected number of analyses performed?

SOLUTION

This example is concerned with two random variables:

X = the number of individuals who have the disease;
Y = the number of tests that must be carried out.

It is easy to see that $X \sim \text{Bi}(10, 0.1)$, but we require to find $E(Y)$ and it is by no means immediately obvious what the probability distribution of Y is. One way to tackle problems like this is to treat Y as a function of the random variable X, and then use the known probability distribution of X to derive the probability distribution of Y. In this case, Y can take just two possible values:

$Y = 1$ if only the composite test is required, that is no one has the disease ($X = 0$);
$Y = 11$ if the composite test is positive and all the individual tests have then to be carried out as well, that is at least one person has the disease ($X \geq 1$).

So, $Y = 1 \Leftrightarrow X = 0$ and $Y = 11 \Leftrightarrow X \geq 1$.

Hence

$$P(Y = 1) = P(X = 0) = (0.9)^{10} = 0.3487$$
$$P(Y = 11) = P(X \geq 1) = 1 - 0.3487 = 0.6513$$

and

$$E(Y) = (1 \times 0.3487) + (11 \times 0.6513) = 7.513$$

In other words, if this procedure were carried out a large number of times, only 7.5 tests would be needed on average, as opposed to ten that would be required if every sample were automatically tested individually. This could be a very useful saving when a large number of samples have to be analysed in a short space of time (e.g. during routine screening or an epidemic).

Here is another example, where Y is specified explicitly as a mathematical function of X. Notice that, again, the probability distribution of Y is found by considering a sequence of equivalent events specified in terms of X.

Example 2

Suppose that X is a discrete random variable with the probability distribution shown in the following table. Find the probability distribution of (a) $Y = -X$, (b) $Y = X^2$.

x	-2	-1	0	1	2
$P(X = x)$	0.10	0.15	0.20	0.25	0.30

SOLUTION

(a)

$$P(Y = 2) = P(X = -2) = 0.10$$
$$P(Y = 1) = P(X = -1) = 0.15$$

etc.

y	-2	-1	0	1	2
$P(Y = y)$	0.30	0.25	0.20	0.15	0.10

(b)

$$P(Y = 4) = P(X = -2 \text{ or } 2) = 0.10 + 0.30 = 0.40$$
$$P(Y = 1) = P(X = -1 \text{ or } 1) = 0.15 + 0.25 = 0.40$$
$$P(Y = 0) = P(X = 0) = 0.20$$

y	0	1	4
$P(Y = y)$	0.20	0.40	0.40

Example 3

A potential customer suggests the following quality assurance procedure to a manufacturer of pumps. Before shipping a lot of 25 pumps, the manufacturer would sample and test five of them at random. The purchaser would accept the lot if no more than one of the pumps tested was defective, though any pump that was tested and found to be defective would have to be repaired. Otherwise, the purchaser would require the manufacturer to test every pump in the lot and repair all that were found to be defective. The manufacturer estimates that it costs £10 to test a pump and £50 to repair one. If two of the pumps in a lot are defective, find the expected cost of complying with the purchaser's quality assurance procedure.

SOLUTION

Let X be the number of defective pumps in the sample. Then, $X \sim \text{Hyp}(5, 25, 2)$. In particular, X has range space $R_X = \{0, 1, 2\}$.

Let Y be the total cost (in £) incurred by the producer. Then if $X = 0$, $Y = 50$ (the cost of five tests),

$$P(Y = 50) = P(X = 0) = \frac{20 \times 19}{25 \times 24} = 0.633$$

if $X = 1$, $Y = 100$ (the cost of five tests and one repair),

$$P(Y = 100) = P(X = 1) = \frac{2 \times 20 \times 5}{25 \times 24} = 0.333$$

if $X = 2$, $Y = 350$ (the cost of 25 tests and two repairs),

$$P(Y = 350) = P(X = 2) = \frac{5 \times 4}{25 \times 24} = 0.033$$

Therefore,

$$E(Y) = (50 \times 0.633) + (100 \times 0.333) + (350 \times 0.033) = 76.5$$

In other words, it would cost the manufacturer £76.50 on average to comply with the proposed quality assurance procedure, if there were two defective pumps in the lot.

EXERCISES ON 14.1

1. The discrete random variable X has the probability distribution given in the following table. Find the probability distribution of Y, where Y is given by: (a) $Y = 1 - X$; (b) $Y = 4X$.

x	0	$\frac{1}{4}$	$\frac{1}{2}$	$\frac{3}{4}$	1
$P(X = x)$	0.25	0.36	0.18	0.12	0.09

2. The discrete random variable X has the probability distribution given in the following table. Define the random variable Y by $Y = X^2$. Find the probability distribution of Y.

x	-2	-1	0	1	2
$P(X = x)$	0.1	0.2	0.4	0.2	0.1

3. In Example 3, find the manufacturer's expected cost of complying with the quality assurance plan when there are three defective pumps in the lot.
4. Inspired by the Citizen's Charter, a public utility is considering the following scheme. Every time the supply of its service is disrupted, the utility will give a rebate of £10 to each customer that is affected, up to a maximum of £20 to any individual in any year. If the annual number of disruptions to service suffered by an individual customer is a Po(0.5) random variable, find the expected cost of this scheme per customer per annum.

14.2 Continuous random variables

Example 4

The opening of a hose has nominal radius 1 cm. Owing to random fluctuations in the manufacturing process, however, the true radius (centimetres) of a hose of this type is a continuous random variable, X, which can be assumed to have a Un(0.9, 1.1) distribution. The rate of flow of fluid through this hose will depend on the cross-sectional area of the opening, $Y = \pi X^2$. How can we find the probability density function of the continuous random variable, Y?

SOLUTION

The key to this question is again to set up a set of equivalent events expressed in terms of both X and Y. There are three steps in the procedure we will adopt: (1) find the range space of Y, R_Y; (2) determine the cumulative distribution function of Y; (3) find the probability density function of Y by differentiating the cumulative distribution function.

In this example, since the range space of X is $R_X = \{x : 0.9 < x < 1.1\}$, then the range space of Y is $R_Y = \{y : 0.9^2\pi < y < 1.1^2\pi\} = \{y : 0.81\pi < y < 1.21\pi\}$. This completes step (1).

The crucial point to note for step (2) is that

$$P(Y \leq y) = P(X \leq x)$$

where $y = \pi x^2$, since these are just two ways of describing the same event. For any value, y, in R_Y, therefore,

$$F_Y(y) = P(Y \leq y) = P(\pi X^2 \leq y) = P\left(X \leq \sqrt{\frac{y}{\pi}}\right) = F_X\left(\sqrt{\frac{y}{\pi}}\right)$$

Since $X \sim \text{Un}(0.9, 1.1)$, then $f_X(x) = [1/(1.1 - 0.9)] = 5$ $(0.9 < x < 1.1)$. So,

$$F_Y(y) = F_X\left(\sqrt{\frac{y}{\pi}}\right) = \int_{0.9}^{\sqrt{y/\pi}} 5\,\mathrm{d}x = 5\left(\sqrt{\frac{y}{\pi}} - 0.9\right)$$

Differentiating with respect to y, we obtain

$$f_Y(y) = \frac{5}{2\sqrt{\pi y}} \qquad 0.81\pi < y < 1.21\pi$$

(Note that this is not a uniform distribution.)

This example introduces by far the most common type of function that is of practical interest, a strictly monotonic function. If $h(x_1) > h(x_2)$ whenever $x_1 > x_2$, then h is a **strictly increasing function**. In a similar way, if $h(x_1) < h(X_2)$ whenever $x_1 > x_2$, then h is a **strictly decreasing function**. Strictly increasing and decreasing functions are both classed as **strictly monotonic functions**. We can find the probability density function of a strictly monotonic function of the random variable X by applying the following general theorem.

• *Theorem 14.1*

Suppose that X is a continuous random variable with range space $R_X = \{x : a < x < b\}$ and with p.d.f. $f_X(x)$, $x \in R_X$. Define the random variable Y by $Y = h(X)$, where h is a strictly increasing, differentiable function on R_X. Then the range space of Y is $R_Y = \{y : h(a) < y < h(b)\}$ and the p.d.f. of Y is

$$f_Y(y) = f_X(h^{-1}(y))\frac{\mathrm{d}x}{\mathrm{d}y}, \quad h(a) < y < h(b)$$

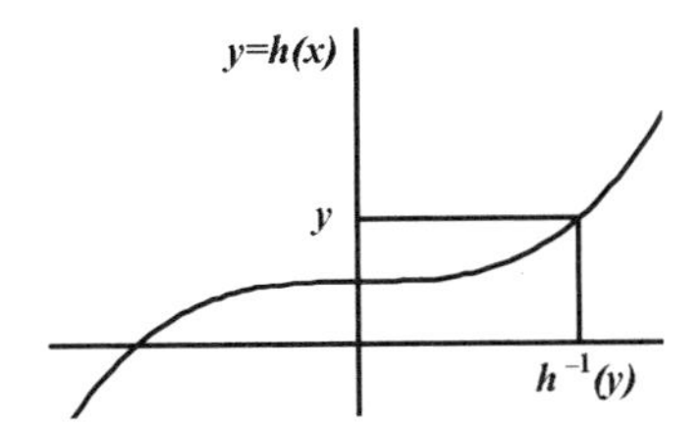

PROOF

It is clear that $R_Y = \{y : h(a) < y < h(b)\}$. For any y in this range,

$$\begin{aligned} F_Y(y) &= P(Y \le y) \\ &= P[h(X) \le y] \\ &= P[X \le h^{-1}(y)] \\ &= F_X(h^{-1}(y)) \end{aligned}$$

Therefore

$$\begin{aligned} f_Y(y) = \frac{\mathrm{d}}{\mathrm{d}y}F_X(h^{-1}(y)) &= \frac{\mathrm{d}}{\mathrm{d}x}F_X(h^{-1}(y))\frac{\mathrm{d}x}{\mathrm{d}y} \quad \text{(chain rule)} \\ &= f_X(h^{-1}(y))\frac{\mathrm{d}x}{\mathrm{d}y} \end{aligned}$$

A similar result holds when $h(X)$ is a strictly decreasing function (see Exercise 3 on p. 127).

• *Example 4 (continued)*

Here, $Y = \pi X^2$; that is, $h(x) = \pi x^2$, which is a strictly increasing function on the range space, $R_X = \{x : 0.9 < x < 1.1\}$. So the range space of Y is $\{y : h(0.9) < y < h(1.1)\}$, that is $\{y : 0.81\pi < y < 1.21\pi\}$.

$y = h(x) = \pi x^2$, so it follows that

$$x = \sqrt{y/\pi}\ (= h^{-1}(y)) \quad \text{and} \quad \frac{dx}{dy} = \frac{1}{2\sqrt{\pi y}}$$

The probability density function of X is $f_X(x) = 5\ (0.9 < x < 1.1)$. Therefore,

$$f_Y(y) = f_X\left(\sqrt{y/\pi}\right)\frac{1}{2\sqrt{\pi y}} = \frac{5}{2\sqrt{\pi y}} \qquad 0.81\pi < y < 1.21\pi$$

Example 5

We have already seen (Chapter 12) that, if $X \sim \text{Ex}(\theta)$, then $Y = kX\ (k > 0)$ has an $\text{Ex}(\theta/k)$ distribution. We can now prove this result using Theorem 14.1.

$h(x) = kx$, which is a strictly increasing function on $R_X = \{x : x > 0\}$. It follows that $R_Y = \{y : y > 0\}$. $y = h(x) = kx$, so

$$x = y/k\ (= h^{-1}(y)) \quad \text{and} \quad \frac{dx}{dy} = \frac{1}{k}$$

Since the probability density function of X is $f_X(x) = \theta e^{-\theta x}\ (x > 0)$, then

$$f_Y(y) = f_X(h^{-1}(y))\frac{dx}{dy} = f_X\left(\frac{y}{k}\right)\frac{1}{k} = \theta e^{-\theta y/k}\frac{1}{k} = \frac{\theta}{k}e^{-\theta y/k} \qquad y > 0$$

and so Y has an $\text{Ex}(\theta/k)$ distribution.

Example 6

Suppose that $X \sim N(0, 1)$ and define the random variable Y by $Y = X^2$. The function $h(x) = x^2$ is not strictly increasing on the range space of X, which is $(-\infty, \infty)$. So, Theorem 14.1 cannot be used to find the probability density function of Y. Instead, we proceed directly, as for Example 4.

$R_Y = \{y : y \geq 0\}$. For any $y \in R_Y$,

$$F_Y(y) = P(Y \leq y) = P(X^2 \leq y) = P(-\sqrt{y} \leq X \leq \sqrt{y}) = F_X(\sqrt{y}) - F_X(-\sqrt{y})$$

So,

$$f_Y(y) = \frac{d}{dy}F_Y(y) = \frac{1}{2\sqrt{y}}f_X(\sqrt{y}) - \frac{-1}{2\sqrt{y}}f_X(-\sqrt{y})$$
$$= \frac{1}{2\sqrt{y}}[f_X(\sqrt{y}) + f_X(-\sqrt{y})]$$

Since the Standard Normal p.d.f., $f_X(x) = (1/\sqrt{2\pi})\exp(-x^2/2)$, is symmetric about 0, then

$$f_Y(y) = \frac{1}{2\sqrt{y}}[2f_X(\sqrt{y})] = \frac{1}{\sqrt{2\pi}\sqrt{y}}e^{-y/2} \qquad y > 0$$

This important distribution is called **the chi-squared distribution** with one degree of freedom, written $Y \sim \chi^2(1)$.

EXERCISES ON 14.2

1. Suppose that $X \sim \mathrm{N}(\mu, \sigma^2)$ and let $Y = aX + b$, for some real constants, a and b. Prove, using Theorem 14.1, that Y is also normally distributed.
2. The radius of a (spherical) ball bearing manufactured by a certain machine is a random variable, X, with probability density function $f_X(x) = 30x^2(1-x)^2$ $(0 < x < 1)$. Find the probability density function of (a) the surface area of a ball bearing, $Y = 4\pi X^2$, and (b) the volume of a ball bearing, $Z = 4\pi X^3/3$.
3. Suppose that X is a continuous random variable with range space $R_X = \{x : a < x < b\}$ and with p.d.f. $f_X(x)$, $x \in R_X$. Define the random variable Y by $Y = h(X)$, where h is a strictly *decreasing*, differentiable function on R_X. Prove that

$$f_Y(y) = f_X(h^{-1}(y))\left(-\frac{\mathrm{d}x}{\mathrm{d}y}\right) \quad h(b) < y < h(a)$$

4. Suppose that the random variable X has range space $R_X = \{x : 0 < x < 1\}$ and p.d.f. $f_X(x) = 6x(1-x)$. Show that $Y = 1 - X$ has the same distribution as X itself. [Use the result proved in Exercise 3 above.]
5. Suppose that the random variable X has probability density function

$$f_X(x) = \frac{1}{\pi(1+t^2)} \qquad -\infty < x < \infty$$

Find the probability density function of $Y = X^2$. [Note: X and Y are examples of, respectively, a t and an F distribution, which are both important in statistics.]

Summary

This chapter has shown how to find the probability distribution or probability density function of a function of a random variable. In each case, the method involves writing a sequence of events in terms of both the original random variable, X, and the derived random variable, $Y = h(X)$. Theorem 14.1 can be used to simplify the calculation when X is continuous and h is a strictly monotonic function. The most important non-monotonic function is $Y = X^2$, when R_X is the set of real numbers; a worked example of this problem has been presented.

FURTHER EXERCISES

1. About 10% of the high-pressure hoses manufactured in a factory have defects that will cause them to leak when used. Every hose is inspected before it leaves the factory, but 2% of defective hoses pass this inspection, while 5% of perfect hoses fail it. Hoses that fail the inspection are scrapped at a cost (or negative profit) to the manufacturer of £10 each. Perfect hoses are sold at a profit of £5. Defective hoses, once sold, will later generate a customer complaint which will lead to them being exchanged, at a total cost to the manufacturer of £20 each. Find the manufacturer's expected profit per hose.

2. A multiple choice examination consists of ten questions. A student who gives the correct answer to a question is awarded two marks, but a student who gives the wrong (or no) answer has one mark deducted. Write down the relationship between the number of questions a student answers correctly (X) and the mark the student gets for the paper (Y). If a student has probability 0.6 of answering each question correctly (independently of all the other questions), find the probability that the student is awarded more than ten marks for the paper.
3. Exercise 4 at the end of Chapter 8 introduced the problems faced by a drunk man requiring to make his way along a straight road to reach home. Each time he moves, he has probability θ of stepping forward 1 metre and probability $1-\theta$ of stepping back 1 metre, independently of what happens every other time he moves. Let X_t denote the overall distance forward (metres) he has moved after t steps. Write down the relationship between X_t and the number of steps forward he takes (Y). Hence find the probability distribution of X_t.
4. Suppose that $X \sim \text{Ex}(\theta)$. If $Y = X^{1/\beta}$, for some positive real constant β, show that Y has the Weibull distribution (Exercise 5 at the end of Chapter 12) $f_Y(y) = \theta\beta y^{\beta-1}\exp(-\theta y^{\beta})$ $(y > 0)$.
5. Suppose that $X \sim \text{N}(\mu, \sigma^2)$. The random variable $Y = \text{e}^X$ is said to have a **Log-normal** distribution. Find its probability density function.
6. Suppose that X is a continuous random variable with range space R_X and cumulative distribution function F_X.
(a) Sketch the curve $y = F_X(x)$ and convince yourself that $P[X \leq F_X^{-1}(y)] = y$ for any y in the range $0 < y < 1$.
(b) Let $Y = F_X(X)$. Show that $Y \sim \text{Un}(0, 1)$.
(c) Let $Z = -\log_e[1 - F_X(x)]$. Show that $Z \sim \text{Ex}(1)$.
7. When a uniform gas is contained under pressure that is constant in all directions, then the speed of a randomly selected gas molecule, S, has probability density function

$$\sqrt{\frac{2}{\pi}}\frac{s^2}{\sigma^3}\exp(-s^2/2\sigma^2) \quad s > 0$$

Find the probability density function of the random variable $E = mS^2/2$, which is the kinetic energy of a randomly selected molecule.
8. A computer is programmed to generate X, a uniformly distributed real number between 0 and 1. The digit Y is then derived according to the rule

$$Y = y \quad \text{if} \quad y \leq 10X < y + 1 \quad (y = 0, 1, \ldots, 9)$$

Find the probability distribution of Y. [Note: this is an example where a discrete random variable is derived from a continuous one. It is impossible to derive a continuous random variable as a simple function of a discrete random variable.]

15 • Bivariate Discrete Distributions

So far, we have concentrated on experiments whose outcomes can be described adequately by a single random variable. Many experiments result in two or more distinct pieces of information being recorded. An experiment might set out to measure two distinct quantities (e.g. height and weight). Or the basic experiment might be repeated a number of times and the outcome recorded each time. Such experiments need to be described by more than one random variable. This chapter presents some of the probability theory required to describe the joint properties of two discrete random variables.

15.1 The joint probability distribution

Example 1

When Sally and Theresa play tennis, the winner of the match is the first of them to win two sets. The following two discrete random variables can be used jointly to record the outcome of one of the girls' completed matches:

$X =$ the number of sets won by Sally;

$Y =$ the number of sets won by Theresa

Sally believes that she has probability θ $(0 < \theta < 1)$ of winning any set against Theresa (independently of what happens in any other set). Suppose she wants to find the probability of an event such as 'the final score is 2–1 for Sally', that is $P(X = 2 \textit{ and } Y = 1)$. Sally wins 2–1 if and only if she wins one of the first two sets and then wins the third (deciding) set. So

$$\begin{aligned} P(X = 2 \textit{ and } Y = 1) &= P(\text{S wins one of the first two sets } \textit{and} \\ &\qquad \text{S wins the third set}) \\ &= P(\text{S wins one of the first two sets}) \cdot P(\text{S wins third set}) \\ &\qquad\qquad\qquad\qquad\qquad\qquad \text{(independence)} \\ &= [2\theta(1-\theta)]\theta \quad \text{(Binomial probabilities)} \\ &= 2\theta^2(1-\theta) \end{aligned}$$

All the probabilities of the form $P(X = x \textit{ and } Y = y)$ are shown in Table 15.1.

In Example 1, (X, Y) must lie in the set $R_{XY} = \{(0, 2), (1, 2), (2, 0), (2, 1)\}$. R_{XY} is known as the **range space** of (X, Y). We can write $p(x, y) = P(X = x \textit{ and } Y = y)$. Clearly, $p(x, y) = 0$ whenever $(x, y) \notin R_{XY}$. The set of triples $\{(x, y, p(x, y)), (x, y) \in R_{XY}\}$ is known as the **joint probability distribution (p.d.)** of X and Y.

Table 15.1 The joint probability distribution, $p(x, y) = P(X = x \text{ and } Y = y)$, for Example 1.

		x		
$p(x,y)$		0	1	2
y	0	0	0	θ^2
	1	0	0	$2\theta^2(1-\theta)$
	2	$(1-\theta)^2$	$2\theta(1-\theta)^2$	0

A valid joint probability distribution for two discrete random variables satisfies the following two conditions:

(a) $0 \le p(x, y) \le 1$, for every $(x, y) \in \mathbf{R}^2$;
(b)

$$\sum\sum_{(x,y)\in R_{XY}} p(x,y) = 1$$

Using the joint probability distribution for Example 1, it is possible to find probabilities such as

$$\begin{aligned} P(\text{both girls win at least one set}) &= p(1,2) + p(2,1) \\ &= 2\theta(1-\theta)^2 + 2\theta^2(1-\theta) \\ &= 2\theta(1-\theta) \end{aligned}$$

Of particular significance are probabilities of the form $P(X = x)$. In Example 1, Table 15.1 shows that:

$$P(X=0) = p(0,2) = (1-\theta)^2;$$
$$P(X=1) = p(1,2) = 2\theta(1-\theta)^2;$$
$$P(X=2) = p(2,0) + p(2,1) = \theta^2 + 2\theta^2(1-\theta) = \theta^2(3-2\theta).$$

In general, $P(X = x)$ is the sum of all the entries in the appropriate *column* of the joint probability distribution:

$$P(X=x) = \sum_{y:(x,y)\in R_{XY}} p(x,y)$$

In Example 1, the values $p_X(x) = P(X = x)$ $(x = 0, 1, 2)$ define a valid probability distribution for X. We refer to it in this context as the **marginal probability distribution** of X. It is precisely the same p.d. that would have been defined had X been the only piece of information recorded from this experiment. The **range space** of X is $R_X = \{0, 1, 2\}$. In a similar way, the *row* sums of the joint probability distribution define the marginal probability distribution for Y. In Example 1:

$$p_Y(0) = P(Y=0) = p(2,0) = \theta^2;$$
$$p_Y(1) = P(Y=1) = p(2,1) = 2\theta^2(1-\theta);$$
$$\begin{aligned} p_Y(2) &= P(Y=2) = p(0,2) + p(1,2) = (1-\theta)^2 + 2\theta(1-\theta)^2 \\ &= (1-\theta)^2(1+2\theta). \end{aligned}$$

Using the joint probability distribution of X and Y, it is possible to define the **expected value** of any real function, $h(X, Y)$, of X and Y as follows:

$$E[h(X, Y)] = \sum\sum_{(x,y)\in R_{XY}} h(x, y) \cdot p(x, y)$$

In particular, defining $h(X, Y) = X$, then

$$E(X) = \sum\sum_{(x,y)\in R_{XY}} xp(x, y) = \sum_{x=0}^{2} x \sum_{y:(x,y)\in R_{XY}}^{2} p(x, y) = \sum_{x=0}^{2} xp_X(x)$$

which agrees with the usual definition of $E(X)$. It can also be shown that $E(X^2)$, and hence $V(X)$, is found from the marginal distribution of X in the usual way. Similarly, $E(Y)$ and $V(Y)$ can be found from the marginal distribution of Y.

Example 2

Suppose that X and Y are discrete random variables with range space $R_{XY} = \{(x, y) : x = -1, 0, 1; y = -1, 0, 1\}$ and joint distribution, $p(x, y)$, given in the following table:

		x		
$p(x, y)$		-1	0	1
y	-1	0.1	0.1	0.1
	0	0.1	0.2	0.1
	1	0.1	0.1	0.1

Using $p(x, y)$, it is possible to calculate probabilities such as:

$$P(X > Y) = p(0, -1) + p(1, -1) + p(1, 0) = 0.1 + 0.1 + 0.1 = 0.3;$$
$$P(X = Y) = p(-1, -1) + p(0, 0) + p(1, 1) = 0.1 + 0.2 + 0.1 = 0.4.$$

Also, the expected value of the product of X and Y is

$$\begin{aligned} E(XY) &= \sum\sum_{(x,y)\in R_{XY}} xy \cdot p(x, y) \\ &= (-1 \times -1 \times 0.1) + (0 \times -1 \times 0.1) + (1 \times -1 \times 0.1) \\ &\quad + (-1 \times 0 \times 0.1) + (0 \times 0 \times 0.2) + (1 \times 0 \times 0.1) \\ &\quad + (-1 \times 1 \times 0.1) + (0 \times 1 \times 0.1) + (1 \times 1 \times 0.1) \\ &= 0.1 + 0 - 0.1 + 0 + 0 + 0 - 0.1 + 0 + 0.1 = 0 \end{aligned}$$

The marginal distribution of X is found by calculating the column sums in this table:

$$p_X(-1) = p(-1, -1) + p(-1, 0) + p(-1, 1) = 0.1 + 0.1 + 0.1 = 0.3;$$
$$p_X(0) = p(0, -1) + p(0, 0) + p(0, 1) = 0.1 + 0.2 + 0.1 = 0.4;$$
$$p_X(1) = p(1, -1) + p(1, 0) + p(1, 1) = 0.1 + 0.1 + 0.1 = 0.3.$$

So, we can find the moments of X from its marginal distribution in the usual way:

$$E(X) = (-1 \times 0.3) + (0 \times 0.4) + (1 \times 0.3) = 0$$
$$E(X^2) = (1 \times 0.3) + (0 \times 0.4) + (1 \times 0.3) = 0.6$$
$$V(X) = E(X^2) - [E(X)]^2 = 0.6$$

In this example, because of the symmetry of $p(x, y)$, Y has the same marginal distribution and moments as X.

Example 3

A medical historian is investigating the way a famous doctor of the early part of the century used three similar diagnoses, A, B and C. The historian has discovered the great man's notes on five relevant cases from a two-year period in his early days, and records X = the number of cases diagnosed A and Y = the number of cases diagnosed B. Since there are only five cases in all, $X + Y \leq 5$. The historian does not need to record the number of cases diagnosed C, since there must be exactly $5 - X - Y$ of them.

The range space of (X, Y) is

$$R_{XY} = \{(x, y) : x = 0, 1, \ldots, 5; y = 0, 1, \ldots, 5 - x\}.$$

Suppose that, in this period, the great man was using the diagnoses A, B and C with respective probabilities θ_1, θ_2 and $1 - \theta_1 - \theta_2$, *independently* in all the cases. Then, any *particular* outcome that gives $X = x$ and $Y = y$ has probability $\theta_1^x \theta_2^y (1 - \theta_1 - \theta_2)^{5-x-y}$. But, there are (Chapter 4),

$$\frac{5!}{x!y!(5 - x - y)!}$$

different ways to choose x cases for diagnosis A, y cases for diagnosis B and $5 - x - y$ cases for diagnosis C. So,

$$p(x, y) = P(X = x \text{ and } Y = y) = \frac{5!}{x!y!(5 - x - y)!} \theta_1^x \theta_2^y (1 - \theta_1 - \theta_2)^{5-x-y}$$
$$(x, y) \in R_{XY}$$

X and Y are said to have a **Trinomial** distribution, written $(X, Y) \sim \text{Tri}(5, \theta_1, \theta_2)$. This is an extension of the Binomial distribution for use when there are three

Table 15.2 The joint probability distribution, $p(x, y)$, when $(X, Y) \sim \text{Tri}(5, 1/3, 1/3)$.

		x					
$p(x, y)$		0	1	2	3	4	5
y	0	1/243	5/243	10/243	10/243	5/243	1/243
	1	5/243	20/243	30/243	20/243	5/243	–
	2	10/243	30/243	30/243	10/243	–	–
	3	10/243	20/243	10/243	–	–	–
	4	5/243	5/243	–	–	–	–
	5	1/243	–	–	–	–	–

possible outcomes of an experiment, rather than two. Table 15.2 shows the joint probability distribution of X and Y when $\theta_1 = \theta_2 = 1/3$.

The range space of X is $R_X = \{0, 1, 2, 3, 4, 5\}$. In the general case, when $(X, Y) \sim \text{Tri}(5, \theta_1, \theta_2)$, for any $x \in R_X$,

$$
\begin{aligned}
p_X(x) &= \sum_{y=0}^{5-x} p(x, y) \quad (\text{since } p(x, y) = 0 \text{ if } x + y \geq 5) \\
&= \sum_{y=0}^{5-x} \frac{5!}{x!y!(5-x-y)!} \theta_1^x \theta_2^y (1 - \theta_1 - \theta_2)^{5-x-y} \\
&= \frac{5!}{x!(5-x)!} \theta_1^x \sum_{y=0}^{5-x} \frac{(5-x)!}{y!(5-x-y)!} \theta_2^y (1 - \theta_1 - \theta_2)^{(5-x)-y} \\
&= \binom{5}{x} \theta_1^x \sum_{y=0}^{5-x} \binom{5-x}{y} \theta_2^y (1 - \theta_1 - \theta_2)^{(5-x)-y} \\
&= \binom{5}{x} \theta_1^x [\theta_2 + (1 - \theta_1 - \theta_2)]^{5-x} \quad (\text{Binomial Theorem}) \\
&= \binom{5}{x} \theta_1^x [1 - \theta_1]^{5-x}
\end{aligned}
$$

So, the marginal distribution of X is the $\text{Bi}(5, \theta_1)$ distribution. In a similar way, it can be shown that (marginally) $Y \sim \text{Bi}(5, \theta_2)$. Hence

$$
\begin{aligned}
E(X) &= 5\theta_1 \quad \text{and} \quad V(X) = 5\theta_1(1 - \theta_1) \\
E(Y) &= 5\theta_2 \quad \text{and} \quad V(Y) = 5\theta_2(1 - \theta_2)
\end{aligned}
$$

In general, X and Y have a Trinomial distribution, $(X, Y) \sim \text{Tri}(n, \theta_1, \theta_2)$, where n is a positive integer, $0 \leq \theta_1, \theta_2 \leq 1$ and $\theta_1 + \theta_2 \leq 1$, if

$$
\begin{aligned}
R_{XY} &= \{(x, y) : x = 0, 1, \ldots, n; y = 0, 1, \ldots, n; x + y \leq n\} \\
p(x, y) &= \frac{n!}{x!y!(n-x-y)!} \theta_1^x \theta_2^y (1 - \theta_1 - \theta_2)^{n-x-y} \quad (x, y) \in R_{XY}
\end{aligned}
$$

When n objects are assigned independently to m categories, and $X_1, \ldots, X_{m-1}$ are the numbers of objects assigned to any $m - 1$ of the categories, then $(X_1, \ldots, X_{m-1})$ are said to have a (joint) **Multinomial** distribution (see more advanced textbooks for details).

EXERCISES ON 15.1

1. Suppose that X and Y have the joint probability distribution shown in the table below.
   (a) Find $P(X > Y)$, $P(X + Y > 0)$, $P(X > 0 \textit{ and } Y > 0)$.
   (b) Find the marginal distributions of X and Y.
   (c) Find $E(2X + 2Y)$.

$p(x, y)$		x			
		$-\frac{1}{2}$	0	$\frac{1}{2}$	1
y	-1	0.1	0	0.1	0.1
	$-\frac{1}{2}$	0.1	0	0.1	0.1
	0	0	0.1	0	0
	$\frac{1}{2}$	0.1	0	0.1	0.1

2. Suppose that X and Y have the joint probability distribution shown in the table below.
(a) Find $P(X = Y)$, $P(XY = 0)$, $P(X + Y \leq 2)$.
(b) Find the marginal distribution of Y. Hence find $P(Y > 0)$.
(c) Find $E(XY)$, and show that $E(XY) - E(X) \cdot E(Y) = 0$.

$p(x, y)$		x		
		0	1	5
y	0	0.16	0.12	0.12
	1	0.12	0.09	0.09
	5	0.12	0.09	0.09

3. Working for the students' union in a large university, you ask ten students independently for their opinion about proposed changes to the library's opening hours. Suppose that a half of all students oppose the proposed change, a quarter support it and the remaining quarter have no preference. Let X and Y (respectively) be the number of students in your sample who oppose and who support the change. Assuming that X and Y follow the appropriate Trinomial distribution, find:
(a) $P(X = 5 \text{ and } Y = 5)$, $P(X = 7 \text{ and } Y = 3)$;
(b) $P(X > 7)$, $E(X)$, $V(X)$;
(c) $P(Y \leq 3)$, $E(Y)$, $V(Y)$.

15.2 Conditional distributions

In Example 1, Y must equal 2 if X equals 0 (since, then, Theresa has won both sets played). This is just an example of a conditional probability:

$$P(Y = 2 \mid X = 0) = 1$$

In general, by the definition of conditional probability (Chapter 6),

$$P(Y = y \mid X = x) = \frac{P(Y = y \cap X = x)}{P(X = x)} = \frac{p(x, y)}{p_X(x)}$$

for $x \in R_X$ and y such that $(x, y) \in R_{XY}$

$$P(X = x \mid Y = y) = \frac{P(X = x \cap Y = y)}{P(Y = y)} = \frac{p(x, y)}{p_Y(y)}$$

for $y \in R_Y$ and x such that $(x, y) \in R_{XY}$

We write $p_{Y|X}(y \mid x) = P(Y = y \mid X = x)$. The set of conditional probabilities $\{(y, p(y \mid x))\}$, for a *fixed* value of x and for *all* y such that $(x, y) \in R_{XY}$, is known as the **conditional probability distribution** of Y **given** $X = x$ (or just given x). In Example 1, the *conditional* probability distribution of Y *given* $X = 2$ is

$$p_{Y|X}(0 \mid 2) = \frac{p(2,0)}{p_X(2)} = \frac{\theta^2}{3\theta^2 - 2\theta^3} = \frac{1}{3 - 2\theta}$$

$$p_{Y|X}(1 \mid 2) = \frac{p(2,1)}{p_X(2)} = \frac{2\theta^2(1-\theta)}{3\theta^2 - 2\theta^3} = \frac{2 - 2\theta}{3 - 2\theta}$$

Since the conditional probability distribution, $p_{Y|X}(y \mid x)$, for *fixed* x, is a well-defined probability distribution, then these probabilities must sum to 1. The *conditional* distribution of X *given* $Y = y$, $p_{X|Y}(x \mid y)$, for *fixed* $y \in R_Y$, is defined in a similar way.

Example 2 (continued)

The conditional probability distribution of Y given that $X = 0$ is

$$p_{Y|X}(-1 \mid 0) = \frac{p(0,-1)}{p_X(0)} = \frac{0.1}{0.4} = \frac{1}{4}$$

$$p_{Y|X}(0 \mid 0) = \frac{p(0,0)}{p_X(0)} = \frac{0.2}{0.4} = \frac{1}{2}$$

$$p_{Y|X}(1 \mid 0) = \frac{p(0,1)}{p_X(0)} = \frac{0.1}{0.4} = \frac{1}{4}$$

The conditional probability distribution of X given that $Y = -1$ is

$$p_{X|Y}(-1 \mid -1) = \frac{p(-1,-1)}{p_Y(-1)} = \frac{0.1}{0.3} = \frac{1}{3}$$

$$p_{X|Y}(0 \mid -1) = \frac{p(0,-1)}{p_Y(-1)} = \frac{0.1}{0.3} = \frac{1}{3}$$

$$p_{X|Y}(1 \mid -1) = \frac{p(1,-1)}{p_Y(-1)} = \frac{0.1}{0.3} = \frac{1}{3}$$

Example 3 (continued)

Given that $X = x$, where $X \in R_X = \{0, 1, 2, 3, 4, 5\}$, then Y must be less than or equal to $5 - x$. The *conditional* probability distribution of Y *given* that $X = x$ is

$$\begin{aligned} p_{Y|X} &= \frac{p(x,y)}{p_X(x)} \\ &= \frac{[5!/x!y!(5-x-y)!]\theta_1^x\theta_2^y(1-\theta_1-\theta_2)^{5-x-y}}{[5!/x!(5-x)!]\theta_1^x(1-\theta_1)^{5-x}} \\ &= \frac{(5-x)!}{y!(5-x-y)!} \frac{\theta_2^y(1-\theta_1-\theta_2)^{5-x-y}}{(1-\theta_1)^{5-x}} \\ &= \binom{5-x}{y}\left(\frac{\theta_2}{1-\theta_1}\right)^y\left(1-\frac{\theta_2}{1-\theta_1}\right)^{(5-x)-y} \qquad (y = 0, 1, \ldots, 5-x) \end{aligned}$$

So, *given* that $X = x$, the *conditional* probability distribution of Y is the $\text{Bi}(5 - x, \theta_2/1 - \theta_1)$ distribution. This makes intuitive sense. If exactly x cases were diagnosed A, then each of the remaining $5 - x$ cases would be diagnosed B (with conditional probability $\theta_2/1 - \theta_1$) or not B (with conditional probability $1 - \theta_2/(1 - \theta_1)$).

EXERCISES ON 15.2

1. For X and Y described in Exercise 1 on pp. 133–4, find the *conditional* probability distribution of:
   (a) X *given* that $Y = \frac{1}{2}$;
   (b) X *given* that $Y = 0$;
   (c) Y *given* that $X = -\frac{1}{2}$.
2. For X and Y described in Exercise 2 on p. 134, find the *conditional* probability distribution of X *given* that (a) $Y = 0$; (b) $Y = 1$; (c) $Y = 5$. Compare these conditional probability distributions with the marginal distribution of X.
3. For X and Y described in Exercise 3 on p. 134, write down the form of the *conditional* probability distribution of:
   (a) X *given* that $Y = 2$;
   (b) Y *given* that $X = 8$.

15.3 Independence and the product model

Example 4

A factory that produces motors for washing machines runs two production lines. Each line is known to produce a defective motor with the same probability, θ. On a particular day, the first line produces m motors and the second line produces n motors. Let the random variables X and Y be the numbers of defective motors produced on the two lines.

For any $x \in \{0, 1, \ldots, m\}$ and any $y \in \{0, 1, \ldots, n\}$, it seems reasonable to assume that the two events $X = x$ and $Y = y$ are independent. So, by the definition of independence (Chapter 6), $P(X = x \text{ and } Y = y) = P(X = x) \cdot P(Y = y)$, that is

$$p(x, y) = p_X(x) \cdot p_Y(y) = \binom{m}{x}\theta^x(1 - \theta)^{m-x}\binom{n}{y}\theta^y(1 - \theta)^{n-y} \quad (x, y) \in R_{XY}$$

where $R_{XY} = \{(x, y) : x = 0, 1, \ldots, m; y = 0, 1, \ldots, n\}$.

In general, the discrete random variables X and Y are said to be **independent** if

$$p(x, y) = p_X(x) \cdot p_Y(y) \quad \text{for all } x \text{ and } y$$

A model in which X and Y are independent is sometimes called a **product model**.

This definition means that the random variables X and Y are independent if and only if

$$p_Y(y) = \frac{p(x, y)}{p_X(x)} = p_{Y|X}(y \mid x)$$

$$\text{for all } x \in R_X \text{ and for all } y \text{ such that } (x, y) \in R_{XY}$$

and

$$p_X(x) = \frac{p(x,y)}{p_Y(y)} = p_{X|Y}(x \mid y)$$

for all $y \in R_Y$ and for all x such that $(x, y) \in R_{XY}$

Example 5

An overworked businessman believes that he has X working lunches in a week, where X is a random variable with probability distribution

x	0	1	2
$p_X(x)$	0.1	0.6	0.3

He also has Y working breakfasts in a week, where Y has probability distribution

y	0	1
$p_Y(y)$	0.8	0.2

Assuming that X and Y are independent random variables, then their joint probability distribution is $p(x, y) = p_X(x) \cdot p_Y(y)$, which is shown in the following table:

			x	
$p(x,y)$		0	1	2
y	0	0.08 $(Z=0)$	0.48 $(Z=1)$	0.24 $(Z=2)$
	1	0.02 $(Z=1)$	0.12 $(Z=2)$	0.06 $(Z=3)$

Let $Z = X + Y$ be the total number of working meals the businessman has in one week. The value of Z associated with each possible combination of X and Y is also shown in the table above. Using $p(x, y)$ we can find the marginal distribution of Z as follows:

$$p_Z(0) = p(0,0) = 0.08;$$
$$p_Z(1) = p(0,1) + p(1,0) = 0.02 + 0.48 = 0.50;$$
$$p_Z(2) = p(1,1) + p(2,0) = 0.12 + 0.24 = 0.36;$$
$$p_Z(3) = p(2,1) = 0.06.$$

z	0	1	2	3
$p_Z(z) = P(Z = z)$	0.08	0.50	0.36	0.06

Example 4 (continued)

We can find the (marginal) probability distribution of the total number of defective motors, $Z = X + Y$, using $p(x, y)$. $R_Z = \{0, 1, \ldots, m+n\}$. For Z to take any particular value, $z \in R_Z$, then Y must equal $z - x$ if X equals x. So,

$$\begin{aligned}
p_Z(z) &= P(Z = z) \\
&= \sum_{x=0}^{m} P(X = x \cap Y = z - x) \\
&= \sum_{x=0}^{m} p(x, z - x) \\
&= \sum_{x=0}^{m} p_X(x) p_Y(z - x) \quad \text{(independence)} \\
&= \sum_{x=0}^{m} \binom{m}{x} \theta^x (1-\theta)^{m-x} \binom{n}{z-x} \theta^{z-x} (1-\theta)^{n-z+x} \\
&= \theta^z (1-\theta)^{m+n-z} \sum_{x=0}^{m} \binom{m}{x} \binom{n}{z-x} \\
&= \binom{m+n}{z} \theta^z (1-\theta)^{m+n-z} \quad \text{(Exercise 4 at the end of Chapter 4)}
\end{aligned}$$

So, $Z = X + Y$ has a $\text{Bi}(m+n, \theta)$ distribution. This makes intuitive sense, since Z can be considered as the number of defectives ('successes') in a total of $m + n$ trials.

Now suppose that the total number of defective motors produced is $Z = z$. Then, X (the number of defective motors produced on the first production line) must be less than or equal to z. The *conditional* probability distribution of X *given* that $Z = z$ can be found as follows:

$$\begin{aligned}
p_{X|Z}(x \mid z) &= P(X = x \mid Z = z) \\
&= \frac{P(X = x \cap Z = z)}{P(Z = z)} \\
&= \frac{P(X = x \cap Y = z - x)}{P(Z = z)} \quad \text{(since } Z = X + Y\text{)} \\
&= \frac{p_X(x) p_Y(z - x)}{p_Z(z)} \quad \text{(independence of } X \text{ and } Y\text{)} \\
&= \frac{\binom{m}{x} \theta^x (1-\theta)^{m-x} \binom{n}{z-x} \theta^{z-x} (1-\theta)^{n-z+x}}{\binom{m+n}{z} \theta^z (1-\theta)^{m+n-z}} \\
&= \frac{\binom{m}{x}\binom{n}{z-x}}{\binom{m+n}{z}} \quad x = 0, 1, \ldots, z
\end{aligned}$$

In other words, *given* that $X + Y = z$, then the *conditional* distribution of X is the $\text{Hyp}(z, m+n, m)$ distribution.

In Example 4, we made use of the *assumption* that X and Y were independent. It is rare to derive a joint probability distribution from other criteria and then discover that X and Y are independent by noticing that $p(x, y) = p_X(x) \cdot p_Y(y)$ for all x and y. The independence of two random variables is usually a model assumption, based on background knowledge about the experiment.

Example 1 (continued)

In Example 1, it is obvious that X and Y are not independent. To show this formally, it is sufficient to find *one* choice of x and *one* choice of y for which $p(x, y)$ is not equal to the product of $p_X(x)$ and $p_Y(y)$. $x = 0$ and $y = 0$ will do, since $p(0, 0)$ is zero, while $p_X(0) = (1 - \theta)^2$ and $p_Y(0) = \theta^2$.

The idea of a product set helps us to spot that some pairs of random variables are not independent. The **Cartesian product** of two sets A and B is the set of pairs $\{(a, b) : a \in A \text{ and } b \in B\}$. For X and Y to be independent, R_{XY} must be the Cartesian product of R_X and R_Y. To see this, consider any $x \in R_X$ and any $y \in R_Y$ such that $(x, y) \notin R_{XY}$. Then, $p(x, y) = 0$, while $p_X(x)$ and $p_Y(y)$ are both greater than 0 by definition.

So, it is immediately obvious that, in Example 3, X and Y are not independent. However, it is possible for R_{XY} to be the Cartesian product of R_X and R_Y, yet for X and Y not to be independent. This is the case in Example 2, where we can see that X and Y are not independent by noticing that $p(-1, -1) = 0.1$ while $p_X(-1) = p_Y(-1) = 0.3$.

Finally, we must briefly consider the joint distribution of two *continuous* random variables, though this topic requires more advanced calculus than is assumed for this book. The **joint cumulative distribution function** of the random variables X and Y is defined by $F(x, y) = P(X \leq x \text{ and } Y \leq y)$.

The **joint probability density function** of the continuous random variables, X and Y, is found from $F(x, y)$ by differentiating once with respect to y and then once with respect to x. This is a second partial derivative, written

$$f(x, y) = \frac{\partial}{\partial x}\frac{\partial}{\partial y} F_{XY}(x, y)$$

X and Y are said to be **independent** if $F(x, y) = F_X(x) \cdot F_Y(y)$ *for all* x and y. Again, this is an extension of the definition of the independence of two events, since $P(X \leq x \text{ and } Y \leq y)$, $P(X \leq x)$ and $P(Y \leq y)$ are all well-defined events. It can be shown that this definition is equivalent to the condition that $f(x, y) = f_X(x) \cdot f_Y(y)$ *for all* x and y.

EXERCISES ON 15.3

1. Show that the random variables X and Y defined in Exercise 1 on pp. 133–4 are not independent. Find the probability distribution of $Z = 2X + 2Y$.
2. For the random variables X and Y defined in Exercise 2 on p. 134, show that $p(x, y) = p_X(x) \cdot p_Y(y)$ for all $(x, y) \in R_{XY}$. Find the probability distribution of $Z = XY$.

Summary

This chapter has defined the joint probability distribution of two discrete random variables, X and Y (say). It has also shown how to derive the (marginal) probability distributions of X, Y and $Z = h(X, Y)$ from the joint probability distribution. The expected value of X, Y or $Z = h(X, Y)$ can be found from the joint probability distribution without first deriving the marginal distribution. The conditional distributions of X given $Y = y$, and of Y given $X = x$, have also been defined. When $p(x, y) = p_X(x) \cdot p_Y(y)$ for all (x, y), then X and Y are said to be independent.

FURTHER EXERCISES

1. The discrete random variables X and Y have the joint probability distribution shown in the following table.
   (a) Find the marginal distribution of X, and hence find $E(X)$ and $V(X)$.
   (b) Find the conditional distribution of X given (i) $Y = -1$; (ii) $Y = 0$; (iii) $Y = 1$.
   (c) Find the marginal distribution of $Z = XY$, and hence find $E(XY) - E(X) \cdot E(Y)$.

			x	
$p(x, y)$		-1	0	1
y	-1	0.05	0.05	0.20
	0	0.05	0.05	0.20
	1	0.20	0.20	0

2. Our family enjoys eating Stocha-sticks, bite-size chocolate biscuits that come in small bags. Based on extensive past experience, we believe that there are either 11, 12 or 13 Stocha-sticks in a bag, with respective probabilities 0.4, 0.5 and 0.1. Assuming that the numbers of biscuits in different bags are independent, write down the joint probability distribution of the numbers of biscuits in two bags. Hence find the (marginal) probability distribution of the total number of biscuits in the two bags. There are four of us; what is the probability that, if we buy two bags of biscuits, the total number of biscuits will be divisible by four?
3. Adam and Brian have just enough money to pay for one turn each on their favourite video game. Whenever a player wins a game, his turn is extended as he is automatically given another game free. Adam and Brian believe that the number of games either of them gets to play, having paid once, is a Ge(0.8) random variable.
   (a) Assuming that the number of games the boys get to play on their two turns are independent, write down a formula for $p(x, y)$.
   (b) Let Z be the total number of games the boys play between them. Show that $P(Z = z) = (z - 1)(0.8)^{z-2}(0.2)^2$ $(z = 2, 3, \ldots)$. Notice that this is the NeBi(2, 0.8) distribution (see Exercise 4 on p. 73).

(c) Suppose that the boys play ten games in total. Find the conditional probability distribution of the number of games played by Adam.

4. To generalize Exercise 3 above, suppose that X and Y are independent $\text{Ge}(\theta)$ random variables. Show that $Z = X + Y \sim \text{NeBi}(2, \theta)$. Show that, *given* $Z = z$, the *conditional* distribution of X is $p(x \mid z) = 1/(z-1)$ $(x = 1, 2, \ldots, z-1)$.

5. A football team believes that, in a home match, the number of goals it scores (X) is a $\text{Po}(\theta_1)$ random variable and the number of goals it concedes (Y) is a $\text{Po}(\theta_2)$ random variable, where X and Y are independent random variables. Let $Z = X + Y$ be the total number of goals scored in a home match.

(a) Write down $p(x, y)$ for $x = 0$, 1, 2 and $y = 0$, 1, 2. Hence find $P(Z = z)$, for $z = 0$, 1 and 2.

(b) Extend your results so far, to show that $Z \sim \text{Po}(\theta_1 + \theta_2)$.

(c) Given that $Z = z$, show that the conditional distribution of X is a Binomial distribution.

16 • Sums of Independent Random Variables

The last chapter showed how to use the joint probability distribution of two discrete random variables to find the probability distribution of their sum (among other simple functions). For example, the sum of two independent binomial random variables (with the same success probability) was proved to be a binomial random variable also. This chapter presents similar results for sums of more than two independent random variables. A different method of proof, based on the moment-generating function, allows the required results to be obtained for continuous as well as discrete random variables.

16.1 The expected value and variance

There are many situations in which we would like to be able to calculate probabilities associated with the sum (or, equivalently, the average) of a sequence of random variables. Here are just a few examples.

(1) The number of items handed into a Lost Property Office in one day is a random variable. We might be interested in the total number of items handed into the office in the course of a week, a month or a year.
(2) The number of defective items produced on a production line in a day is a random variable. The total number of defective items produced by all the production lines in a factory in a day, or the total number of defective items produced by a given production line in a week, is the sum of a sequence of random variables.
(3) A physicist who has developed a method for measuring the speed of light knows that the method is subject to measurement error. The physicist repeats the measurement, which is a random variable, on a number of occasions and compares the average result with the known true value.
(4) The amount of petrol sold in a week by a filling station is a random variable. The total amount of petrol sold by all the filling stations owned by the same company has an important impact on distribution and purchasing decisions.

This section will show how the expected value and variance of the sum of a sequence of random variables relate to the expected values and variances of the original random variables themselves. In each case, a result will be proved for the sum of two discrete random variables and then extended to the sum of any finite number of random variables. The same results hold for sums of continuous random variables too, though we cannot prove this without more advanced calculus.

If $X_1, X_2, \ldots, X_m$ $(m \geq 2)$ are discrete random variables, then their **joint probability distribution** is

$$p(x_1, x_2, \ldots, x_m) = P(X_1 = x_1 \textit{ and } X_2 = x_2 \textit{ and} \ldots \textit{and } X_m = x_m)$$

The discrete random variables $X_1, X_2, \ldots, X_m$ are said to be **independent** if

$$p(x_1, x_2, \ldots, x_m) = p_1(x_1) \cdot p_2(x_2) \cdot \ldots \cdot p_m(x_m) \quad \text{for all } x_1, x_2, \ldots, x_m$$

where $p_i(x_i)$ is the (marginal) probability distribution of X_i.

If $X_1, X_2, \ldots, X_m$ $(m \geq 2)$ are continuous random variables, then their **joint cumulative distribution function** is

$$F(x_1, x_2, \ldots, x_m) = P(X_1 \leq x_1 \textit{ and } X_2 \leq x_2 \textit{ and} \ldots \textit{and } X_m \leq x_m)$$

The continuous random variables $X_1, X_2, \ldots, X_m$ are said to be **independent** if

$$F(x_1, x_2, \ldots, x_m) = F_1(x_1) \cdot F_2(x_2) \cdot \ldots \cdot F_m(x_m) \quad \text{for all } x_1, x_2, \ldots, x_m$$

where $F_i(x_i)$ is the (marginal) cumulative distribution function of X_i.

Result 16.1

For any random variables X and Y, $E(a_1X + a_2Y + b) = a_1E(X) + a_2E(Y) + b$.

PROOF

By the definition of the expected value of a real-valued function of X and Y (page 131),

$$\begin{aligned} E(a_1X + a_2Y + b) &= \sum\sum_{(x,y)\in R_{XY}} (a_1x + a_2y + b)p(x,y) \\ &= a_1 \cdot \sum\sum_{(x,y)\in R_{XY}} x \cdot p(x,y) + a_2 \cdot \sum\sum_{(x,y)\in R_{XY}} y \cdot p(x,y) \\ &\quad + b \cdot \sum\sum_{(x,y)\in R_{XY}} p(x,y) \\ &= a_1 \cdot E(X) + a_2 \cdot E(Y) + b \end{aligned}$$

This result can be extended to any (finite) number of random variables, as follows:

$$E(a_1X_1 + a_2X_2 + \ldots + a_mX_m + b) = a_1E(X_1) + a_2E(X_2) + \ldots + a_mE(X_m) + b$$

Example 1

Suppose that $X_1, X_2, \ldots, X_m$ $(m \geq 2)$ are independent random variables with the same distribution. We say that they are **independent and identically distributed (i.i.d.)** random variables. In particular, suppose that each X_i has expected value μ and variance σ^2. Let

$$\bar{X} = \frac{1}{m}(X_1 + X_2 + \ldots + X_m)$$

be the arithmetic average (mean) of the m random variables. Then,

$$E(\bar{X}) = \frac{1}{m}E(X_1) + \frac{1}{m}E(X_2) + \ldots + \frac{1}{m}E(X_m) = \frac{1}{m}\mu + \frac{1}{m}\mu + \ldots + \frac{1}{m}\mu = \mu$$

The next result, though it does not deal directly with sums of random variables, is required to prove other results that we will meet later.

Result 16.2

Suppose that X and Y are *independent* random variables, and let $g(X)$ and $h(Y)$ be arbitrary real-valued functions of X and Y, respectively. Then $E[g(X) \cdot h(Y)] = E[g(X)] \cdot E[h(Y)]$.

PROOF

Since X and Y are independent, R_{XY} is the Cartesian product of R_X and R_Y.

$$\begin{aligned} E[g(X) \cdot h(Y)] &= \sum_{(x,y) \in R_{XY}}\sum g(x) \cdot h(y) \cdot p(x, y) \\ &= \sum_{x \in R_X}\sum_{y \in R_Y} g(x) \cdot h(y) \cdot p_X(x) \cdot p_Y(y) \quad \text{(independence)} \\ &= \sum_{x \in R_X} g(x) \cdot p_X(x) \sum_{y \in R_Y} h(y) \cdot p_Y(y) \\ &= E[g(X)] \cdot E[h(Y)] \end{aligned}$$

This result too can be extended to any (finite) number of random variables. Suppose that $X_1, X_2, \ldots, X_m$ are *independent* random variables. Then

$$E[g_1(X_1) \cdot g_2(X_2) \cdot \ldots \cdot g_m(X_m)] = E[g_1(X_1)] \cdot E[g_2(X_2)] \cdot \ldots \cdot E[g_m(X_m)]$$

Example 2

The value $E(XY) - E(X)E(Y)$ is known as the **covariance** of X and Y, and is denoted $\text{cov}(X, Y)$. Suppose that X and Y are independent. Then

$$E(XY) = E(X)E(Y)$$

that is,

$$\text{cov}(X, Y) = E(XY) - E(X)E(Y) = 0$$

Notice that the converse of this result does not hold, i.e. $\text{cov}(X, Y) = 0$ does not necessarily imply that X and Y are independent.

Result 16.3

Suppose that X and Y are *independent* random variables. Then $V(a_1X + a_2Y + b) = a_1^2V(X) + a_2^2V(Y)$.

PROOF

Using Results 16.1 and 16.2,

$$\begin{aligned} E[(a_1X + a_2Y + b)^2] &= E[a_1^2X^2 + a_2^2Y^2 + b^2 + 2a_1a_2XY + 2a_1Xb + 2a_2Yb] \\ &= a_1^2E(X^2) + a_2^2E(Y^2) + b^2 + 2a_1a_2E(X)E(Y) \\ &\quad + 2a_1bE(X) + 2a_2bE(Y) \\ [E(a_1X + a_2Y + b)]^2 &= [a_1E(X) + a_2E(Y) + b]^2 \\ &= a_1^2[E(X)]^2 + a_2^2[E(Y)]^2 + b^2 + 2a_1a_2E(X)E(Y) \\ &\quad + 2a_1bE(X) + 2a_2bE(Y) \end{aligned}$$

So,

$$\begin{aligned} V(a_1X + a_2Y + b) &= E[(a_1X + a_2Y + b)^2] - [E(a_1X + a_2Y + b)]^2 \\ &= a_1^2\{E(X^2) - [E(X)]^2\} + a_2^2\{E(Y^2) - [E(Y)]^2\} \\ &= a_1^2V(X) + a_2^2V(Y) \end{aligned}$$

This result too can be extended to more than two variables. Suppose that $X_1, X_2, \ldots, X_m$ are *independent* random variables. Then

$$V(a_1X_1 + a_2X_2 + \ldots + a_mX_m + b) = a_1^2V(X_1) + a_2^2V(X_2) + \ldots + a_m^2V(X_m)$$

Example 1 (continued)

The variance of the mean of m i.i.d. random variables is

$$\begin{aligned} V(\bar{X}) &= \left(\frac{1}{m}\right)^2 V(X_1) + \left(\frac{1}{m}\right)^2 V(X_2) + \ldots + \left(\frac{1}{m}\right)^2 V(X_m) \\ &= \left(\frac{1}{m}\right)^2 \sigma^2 + \left(\frac{1}{m}\right)^2 \sigma^2 + \ldots + \left(\frac{1}{m}\right)^2 \sigma^2 \\ &= \frac{\sigma^2}{m} \end{aligned}$$

Example 3

Suppose that X and Y are i.i.d. random variables, each with expected value μ and variance σ^2. Then

$$E(X - Y) = 1 \cdot E(X) + (-1) \cdot E(Y) = E(X) - E(Y) = 0$$
$$V(X - Y) = 1^2 \cdot V(X) + (-1)^2 \cdot V(Y) = V(X) + V(Y) = 2\sigma^2$$

Example 4

Suppose that $X_1, X_2, \ldots, X_m$ are *independent* random variables, and that each $X_i \sim \text{Ex}(\theta t_i)$, for some parameter θ, where $t_1, t_2, \ldots, t_m$ are positive real constants. Find the expected value and variance of

$$Y = \frac{1}{m}(X_1t_1 + X_2t_2 + \ldots + X_mt_m)$$

SOLUTION

Since $X_i \sim \text{Ex}(\theta t_i)$, then

$$E(X_i) = \frac{1}{\theta t_i} \quad \text{and} \quad V(X_i) = \frac{1}{(\theta t_i)^2}$$

Using Results 16.1 and 16.3,

$$\begin{aligned}
E(Y) &= \frac{t_1}{m}E(X_1) + \frac{t_2}{m}E(X_2) + \ldots + \frac{t_m}{m}E(X_m) \\
&= \frac{t_1}{m}\frac{1}{\theta t_1} + \frac{t_2}{m}\frac{1}{\theta t_2} + \ldots + \frac{t_m}{m}\frac{1}{\theta t_m} \\
&= \frac{1}{m\theta} + \frac{1}{m\theta} + \ldots + \frac{1}{m\theta} \\
&= \frac{1}{\theta} \\
V(Y) &= \left(\frac{t_1}{m}\right)^2 V(X_1) + \left(\frac{t_2}{m}\right)^2 V(X_2) + \ldots + \left(\frac{t_m}{m}\right)^2 V(X_m) \\
&= \left(\frac{t_1}{m}\right)^2 \frac{1}{\theta^2 t_1^2} + \left(\frac{t_2}{m}\right)^2 \frac{1}{\theta^2 t_2^2} + \ldots + \left(\frac{t_m}{m}\right)^2 \frac{1}{\theta^2 t_m^2} \\
&= \frac{1}{m^2\theta^2} + \frac{1}{m^2\theta^2} + \ldots + \frac{1}{m^2\theta^2} \\
&= \frac{1}{m\theta^2}
\end{aligned}$$

EXERCISES ON 16.1

1. Suppose that X, Y and Z are independent random variables, with $E(X) = E(Y) = 2$; $V(X) = V(Y) = 1$; $E(Z) = 5$; $V(Z) = 2$. Find:
   (a) $E(X - Y)$, $V(X - Y)$;
   (b) $E(X - Y + 1)$, $V(X - Y + 1)$;
   (c) $E(2Z)$, $V(2Z)$;
   (d) $E(X + 4Y - 2Z)$, $V(X + 4Y - 2Z)$.
2. The daily number of new customers who telephone a computer consultancy service is a Po(2) random variable. Making any necessary assumptions, find the expected value and variance of the total number of new customers who telephone in the course of a working week of five days.
3. Suppose that $X_1, X_2, \ldots, X_m$ are *independent* random variables, and that each $X_i \sim \text{Po}(\theta t_i)$, for some parameter θ, where $t_1, t_2, \ldots, t_m$ are positive real constants. Find the expected value and variance of:

   (a) $Y = \dfrac{X_1 + X_2 + \ldots X_m}{t_1 + t_2 + \ldots + t_m}$ (b) $Z = \dfrac{1}{m}\left(\dfrac{X_1}{t_1} + \dfrac{X_2}{t_2} + \ldots + \dfrac{X_m}{t_m}\right)$

16.2 Reproductive properties

(The proofs of the results in this section make use of the moment-generating function. Though the results themselves should be studied carefully, the proofs may be ignored on a first reading.)

While it can be useful to know the expected value and the variance of the sum of a sequence of random variables, it is even more useful to know its probability distribution or its probability density function. For some of the standard types of random variable that we have discussed before, the sum of a sequence of random variables has a similar distribution to that of the individual random variables. These results are known as **reproductive properties**.

For example, if X_1 and X_2 are independent random variables, with $X_i \sim \text{Bi}(n_i, \theta)$, $i = 1, 2$, then $X_1 + X_2 \sim \text{Bi}(n_1 + n_2, \theta)$, as proved in Chapter 15. This result can be extended by a process of induction to a sequence $X_1, X_2, \ldots, X_m$ of independent random variables such that $X_i \sim \text{Bi}(n_i, \theta)$, $i = 1, 2, \ldots, m$. Then $X_1 + X_2 + \ldots + X_m \sim \text{Bi}(n_1 + n_2 + \ldots + n_m, \theta)$.

There is another, more straightforward, way to prove results like this, that is based on the moment-generating function. It makes use of the following result.

Result 16.4

Suppose that $X_1, X_2, \ldots, X_m$ are *independent* random variables, and that the moment-generating function of X_i is $M_i(u)$, $i = 1, 2, \ldots, m$. Then the moment-generating function of the random variable $Y = a_1X_1 + a_2X_2 + \ldots + a_mX_m$ is

$$M_Y(u) = M_1(a_1u) \cdot M_2(a_2u) \cdot \ldots \cdot M_m(a_mu)$$

PROOF

$$\begin{aligned} M_Y(u) = E(e^{Yu}) &= E\{\exp[(a_1X_1 + a_2X_2 + \ldots + a_mX_m)u]\} \\ &= E[\exp(X_1a_1u)\exp(X_2a_2u)\ldots\exp(X_ma_mu)] \\ &= E[\exp(X_1a_1u)]E[\exp(X_2a_2u)]\ldots E[\exp(X_ma_mu)] \quad \text{(Result 16.2)} \\ &= M_1(a_1u) \cdot M_2(a_2u) \cdot \ldots \cdot M_m(a_mu) \end{aligned}$$

Result 16.5

Suppose that $Y = X_1 + X_2 + \ldots + X_m$, where $X_1, X_2, \ldots, X_m$ are independent random variables and $X_i \sim \text{Bi}(n_i, \theta)$, $i = 1, 2, \ldots, m$. Then

$$Y \sim \text{Bi}(n_1 + n_2 + \ldots + n_m, \theta).$$

PROOF

Each X_i has moment-generating function $M_i(u) = (\theta e^u + 1 - \theta)^{n_i}$ (Chapter 11, Example 3). So, Y has moment-generating function

$$\begin{aligned} M_Y(u) &= (\theta e^u + 1 - \theta)^{n_1}(\theta e^u + 1 - \theta)^{n_2}\ldots(\theta e^u + 1 - \theta)^{n_m} \\ &= (\theta e^u + 1 - \theta)^{n_1 + n_2 + \ldots + n_m} \end{aligned}$$

which is the moment-generating function of a $\text{Bi}(n_1 + n_2 + \ldots + n_m, \theta)$ random variable. By the uniqueness property of moment-generating functions (Chapter 12), it follows that $X_1 + X_2 + \ldots + X_m \sim \text{Bi}(n_1 + n_2 + \ldots + n_m, \theta)$.

Result 16.6

Suppose that $Y = a_1X_1 + a_2X_2 + \ldots + a_mX_m$, where $X_1, X_2, \ldots, X_m$ are independent random variables and $X_i \sim \text{N}(\mu_i, \sigma_i^2)$, $i = 1, 2, \ldots, m$, and $a_1, a_2, \ldots, a_m$ are real constants. Then $Y \sim \text{N}(a_1\mu_1 + \ldots + a_m\mu_m, a_1^2\sigma_1^2 + \ldots + a_m^2\sigma_m^2)$.

PROOF

Each X_i has moment-generating function $M_i(u) = \exp(u\mu_i + \frac{1}{2}u^2\sigma_i^2)$ (Chapter 13, p. 116). So, Y has moment-generating function

$$\begin{aligned} M_Y(u) &= \exp(a_1u\mu_1 + \tfrac{1}{2}a_1^2u^2\sigma_1^2) \times \exp(a_2u\mu_2 + \tfrac{1}{2}a_2^2u^2\sigma_2^2) \ldots \\ &\quad \times \exp(a_mu\mu_m + \tfrac{1}{2}a_m^2u^2\sigma_m^2) \\ &= \exp[(a_1u\mu_1 + \tfrac{1}{2}a_1^2u^3\sigma_1^3) + (a_2u\mu_2 + \tfrac{1}{2}a_2^2u^2\sigma_2^2) + \ldots \\ &\quad + (a_mu\mu_m + \tfrac{1}{2}a_m^2u^2\sigma_m^2)] \\ &= \exp[u(a_1\mu_1 + a_2\mu_2 + \ldots + a_m\mu_m) + \tfrac{1}{2}u^2(a_1^2\sigma_1^2 + a_2^2\sigma_2^2 + \ldots + a_m^2\sigma_m^2)] \end{aligned}$$

which is the moment-generating function of a Normal distribution with expected value $(a_1\mu_1 + a_2\mu_2 + \ldots + a_m\mu_m)$ and variance $(a_1^2\sigma_1^2 + a_2^2\sigma_2^2 + \ldots + a_m^2\sigma_m^2)$. By the uniqueness property of moment-generating functions, then,

$$Y \sim \text{N}(a_1\mu_1 + a_2\mu_2 + \ldots + a_m\mu_m, a_1^2\sigma_1^2 + a_2^2\sigma_2^2 + \ldots + a_m^2\sigma_m^2)$$

Example 5

The weight (lbs) of a 'quarter-pound' hamburger is an $\text{N}(0.27, 0.0002)$ random variable. Find the probability that the two hamburgers in a 'double' weigh at least half a pound in total.

SOLUTION

Let X and Y be the weights of the individual hamburgers. Then the total weight, T, is an $\text{N}(0.54, 0.0004)$ random variable. So,

$$\begin{aligned} P(T > 0.50) &= P\left(\frac{T - 0.54}{0.02} > \frac{0.50 - 0.54}{0.02} = -2\right) \\ &= 1 - \Phi(-2) = \Phi(2) = 0.9772 \end{aligned}$$

Example 6

The heights, $X_1, X_2, \ldots, X_m$, of m girls are measured on their first day at school. These measurements are a sequence of i.i.d. $\text{N}(\mu, \sigma^2)$ random variables, where μ and σ^2 are the expected value and variance of the heights of all British girls on their first day at school. μ is unknown, and we want to find plausible values for it on the basis of this sample. This is an example of statistical inference, an attempt to draw conclusions about a population from a sample.

In accordance with various statistical criteria and common sense, we decide to estimate the population mean, μ, by the sample mean

$$\bar{X} = \frac{1}{m}\sum_{i=1}^{m} X_i$$

Result 16.6 means that (see Exercise 1 below):

$$\bar{X} \sim \mathrm{N}(\mu, \sigma^2/m)$$

that is,

$$\frac{\bar{X} - \mu}{\sqrt{\sigma^2/m}} \sim \mathrm{N}(0, 1) \quad \text{(see Chapter 13)}$$

and so there is probability 0.95 that the interval $(-1.96, 1.96)$ contains the value

$$\frac{\bar{X} - \mu}{\sqrt{\sigma^2/m}}$$

that is, that the interval $(\bar{X} - 1.96\sqrt{\sigma^2/m}, \bar{X} + 1.96\sqrt{\sigma^2/m})$ contains the value μ.

Even when expressed in this way, this is still a probability statement about the random variable $\bar{X}$. Although μ is unknown, it is not a random variable, but a constant real value.

EXERCISES ON 16.2

1. Suppose that $X_1, X_2, \ldots, X_m$ are i.i.d. $\mathrm{N}(\mu, \sigma^2)$ random variables, and let $Y = X_1 + X_2 + \ldots + X_m$. Use Result 16.6 to write down the distribution of $\bar{X}$ and of Y.
2. The radius (centimetres) of a cog wheel is an $\mathrm{N}(2, 0.0004)$ random variable. The wheels are produced independently and then paired. A pair is satisfactory if the radii of the two wheels differ by no more than 0.03 cm. Find the probability that a pair of wheels is not satisfactory.

Summary

This chapter has shown that the expected value of the sum of a sequence of random variables is equal to the sum of the expected values of the individual random variables. When the random variables are independent, then the variance of the sum is just the sum of the variances. More general results have been presented to deal with the case of a weighted sum. Again in the cases of a sequence of independent random variables, it has been shown that the moment-generating function of a weighted sum of the random variables can be found from the individual moment-generating function. Using this result, a number of reproductive properties have been derived for commonly occurring types of random variable.

FURTHER EXERCISES

(Exercises below that make use of the moment-generating function are marked with a *.)

1. Suppose that $X_1, X_2, \ldots, X_m$ $(m \geq 2)$ are i.i.d. random variables, each with expected value μ and variance σ^2. Find the expected value and the variance of the sum of the m random variables.

*2. Suppose that $Y = X_1 + X_2 + \ldots + X_m$, where $X_1, X_2, \ldots X_m$ are independent random variables and each $X_i \sim \text{Po}(\theta_i)$. Show that Y is also a Poisson random variable. Use this result to find the probability that fewer than five new customers contact the computer consultancy described in Exercise 2 on p. 146 during a working week of five days.

*3. In Exercise 7 at the end of Chapter 12, it was shown that a $\text{Ga}(n, \theta)$ random variable has moment-generating function

$$M_X(u) = \left(\frac{\theta}{\theta - u}\right)^n \quad u < \theta$$

The $\text{Ga}(1, \theta)$ distribution is just another name for the $\text{Ex}(\theta)$ distribution.
(a) Show that, if $X_1, X_2, \ldots, X_m$ are independent random variables and each $X_i \sim \text{Ex}(\theta)$, then $Y = X_1 + X_2 + \ldots + X_m \sim \text{Ga}(m, \theta)$.
(b) Show that, if $X_1, X_2, \ldots, X_m$ are independent random variables and each $X_i \sim \text{Ex}(\theta t_i)$, then $Y = t_1 X_1 + t_2 X_2 + \ldots + t_m X_m \sim \text{Ga}(m, \theta)$.

17 • The Central Limit Theorem

In the last chapter, we derived the exact distributions of sums of some sequences of random variables. It is not always possible to find an exact result in this kind of problem, but it is still of interest to discuss approximate probabilities associated with the sum and average of the random variables, rather than just their expected value and variance. One of the most important theorems in probability, known as the Central Limit Theorem, allows us to do this for an arbitrary sequence of random variables.

17.1 Approximate results

Example 1

Suppose that a computer is programmed to round the costs of large financial transactions to the nearest whole pound (£). If the same institutional client makes 1200 transactions in one year, what is the probability that the total cost, as computed in this way, will differ by less than £5 from the true cost?

Let X_i be the difference (in £) between the computed cost of the ith transaction and its true cost $(i = 1, 2, \ldots, 1200)$. Then we may assume that each $X_i \sim \text{Un}(-0.5, 0.5)$. So the difference between the total cost of the 1200 transactions and the computed cost is

$$Y = X_1 + X_2 + \ldots + X_{1200}$$

To answer the problem posed, we would like to know the distribution of Y. There is no reproductive property that we can call on to find this distribution exactly, but we can make use of the following approximate result.

• *Theorem 17.1 (The Central Limit Theorem)*

Let $X_1, X_2, \ldots, X_n$ be a sequence of independent and identically distributed random variables, each with (finite) expected value μ and (finite) variance σ^2. Then, for sufficiently large n,

$$\frac{\sum_{i=1}^{n} X_i - n\mu}{\sqrt{n\sigma^2}} \approx \text{N}(0, 1)$$

in the sense that

$$\lim_{n\to\infty} \left\{ \frac{\sum_{i=1}^{n} X_i - n\mu}{\sqrt{n\sigma^2}} \leq z \right\} = \Phi(z) \quad \text{for every real number, } z$$

PROOF
Omitted.

We can interpret this result in two equivalent ways:

(1) $\sum_{i=1}^{n} X_i \approx \mathrm{N}(n\mu, n\sigma^2)$ for 'large' n;

(2) $\bar{X} \approx \mathrm{N}(\mu, \sigma^2/n)$ for 'large' n.

The expected values and variances given for the sum and average of the random variables in the above theorem are exactly correct (see Chapter 16); the approximate normality is an important new result.

The Central Limit Theorem (CLT) is an extremely powerful result, since it requires so few assumptions to be made about the probability distribution of the X_i. It is even possible to prove a version of the CLT in which the random variables need not be independent or identically distributed!

Example I (continued)

Using the results derived in Chapter 12, $\mu = E(X_i) = 0$ and $\sigma^2 = V(X_i) = 1/12$. So, using the CLT,

$$\frac{Y - 0}{10} \approx \mathrm{N}(0, 1)$$

Therefore, the probability that the total error is no more than £5 (in either direction) is

$$\begin{aligned} P(-5 < Y < 5) = P(-0.5 < Y/10 < 0.5) &= \Phi(0.5) - \Phi(-0.5) \\ &= 2\Phi(0.5) - 1 = 0.3830 \end{aligned}$$

In the statement of the Central Limit Theorem, it is claimed that the Normal approximation is reasonable if n, the number of random variables that are summed or averaged, is 'sufficiently large'. For most types of distribution, n need only be between 20 and 30 for a good approximation. Suppose, though, that each X_i has an Ex(1) distribution. The following simulation study shows that a rather larger n is required in this case.

A computer was programmed to generate artificial values from the Ex(1) distribution. n of these values were then combined to simulate a value for

$$S(n) = \frac{\sum_{i=1}^{n} X_i - n}{\sqrt{n}}$$

the standardized variable whose distribution, according to the CLT, should tend to the Standard Normal as $n \to \infty$. Figure 17.1 shows the results for 1000 independent simulations when (a) $n = 1$, (b) $n = 10$, (c) $n = 50$, (d) $n = 100$. The histograms give an impression of the shape of the corresponding probability density functions.

For $n = 1$, the histogram is very skewed, reflecting the skewness of the Ex(1) distribution (see Chapter 12). Even for $n = 10$, this skewness persists. By the time we get to $n = 50$, the distribution looks much more symmetric (as it must be for a Standard Normal distribution), and for $n = 100$ the approximation to the Standard Normal seems quite acceptable.

So, for the Ex(1) distribution, it seems to be necessary to sum or average between 50 and 100 independent random variables to get a good Normal approximation. It should be remembered, though, that this a notably poor case,

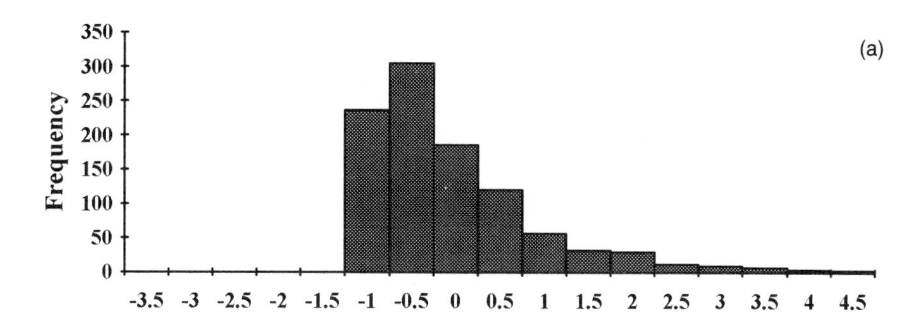

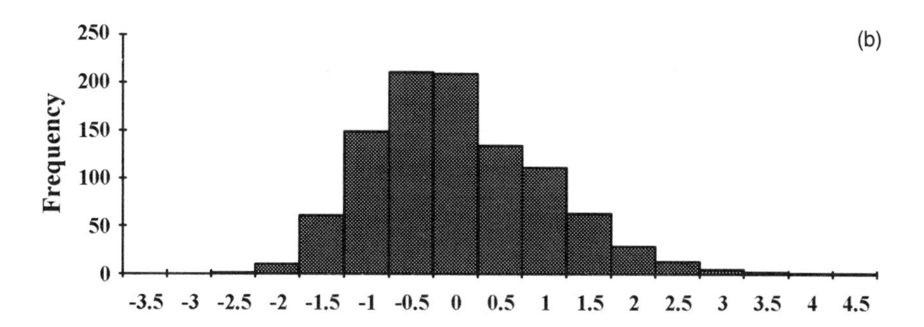

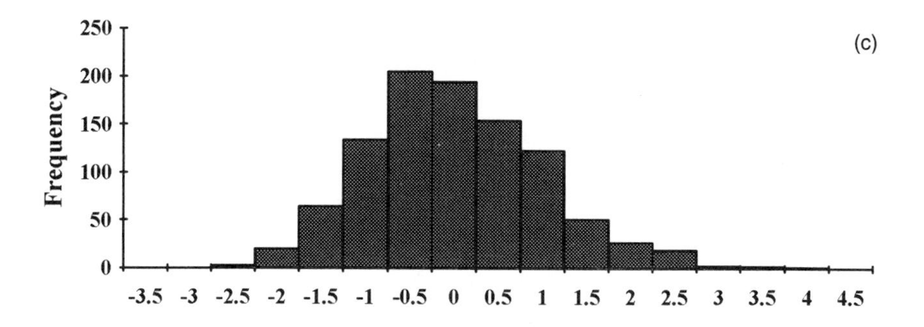

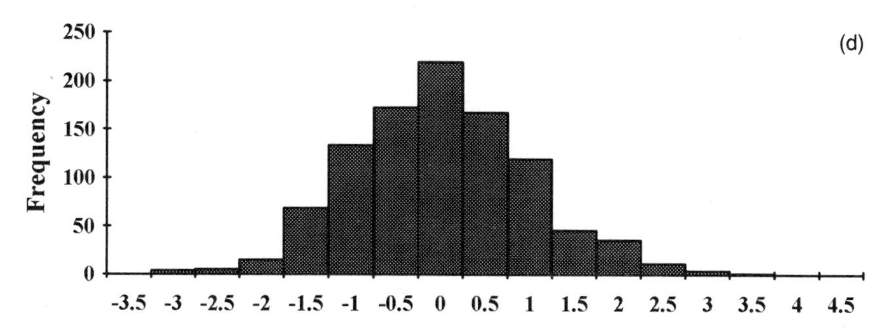

Fig 17.1 1000 simulations of $S(n)$ for (a) $n = 1$, (b) $n = 10$, (c) $n = 50$, (d) $n = 100$.

and that for most other common distributions a much smaller value of n will suffice.

Example 2

A research worker intends to draw a sample of size N from a large population. The researcher will measure for each item in the sample a characteristic that has expected value μ and variance σ^2 in the population, and intends to estimate μ by the sample mean. How large must N be, to give a probability of at least 0.95 that the sample mean differs from μ by at most 0.1σ?

SOLUTION

Let $X_1, X_2, \ldots, X_N$ be the N measurements. These are i.i.d. random variables with $E(X_i) = \mu$ and $V(X_i) = \sigma^2$ $(i = 1, 2, \ldots, N)$. So,

$$Z = \frac{\bar{X} - \mu}{\sqrt{\sigma^2/N}} \approx \mathrm{N}(0, 1)$$

We require to find N such that

$$P\{-0.1\sigma < \bar{X} - \mu < 0.1\sigma\} \geq 0.95$$

that is,

$$P\left\{\frac{-0.1\sigma}{\sqrt{\sigma^2/N}} < Z < \frac{0.1\sigma}{\sqrt{\sigma^2/N}}\right\} \geq 0.95$$

$$P\{-0.1\sqrt{N} < Z < 0.1\sqrt{N}\} \geq 0.95$$

$$2\Phi(0.1\sqrt{N}) - 1 \geq 0.95$$

$$\Phi(0.1\sqrt{N}) \geq 0.975 = \Phi(1.96) \quad \text{(Appendix 2)}$$

$$0.1\sqrt{N} \geq 1.96$$

$$N \geq 384.16$$

So, the smallest acceptable sample size is $N = 385$. Notice that this answer does not depend on μ, σ^2 or the probability distribution of the measured characteristic.

EXERCISES ON 17.1

1. A research and development team have designed a new car gearbox. Prior to releasing the gearbox for sale, they require to conduct final tests on ten of the gearboxes on the firm's accelerated test facility. The gearboxes are to be tested consecutively, one at a time, until failure. Once a gearbox fails, its place on the test facility will be taken immediately by another gearbox. Previous experience suggests that the lifetime (hours) of a new gearbox on this facility is an Ex(0.1) random variable. Find the shortest total time the team must book for the sequence of tests, to have a probability of at least 0.99 of completing all ten tests within the booked time.

17.2 Approximations to the Binomial and Poisson distributions

Example 3

When applied at a given concentration, a pesticide is claimed to kill a fruitfly with probability 0.98. Suppose that, in a field trial, the pesticide is applied to 10,000 fruitflies. What is the probability that at least 9850 of the flies are killed?

SOLUTION

Let X be the number of fruitflies killed in this trial. Then, $X \sim \mathrm{Bi}(10{,}000, 0.98)$, assuming that flies are killed or not killed independently of one another. We require to find $P(X > 9850)$, which is a difficult calculation to perform directly from the probability distribution of the binomial random variable.

The Normal distribution is commonly used to evaluate approximate 'tail' probabilities like this from the Binomial distribution when n is large. This approximation is justified by the CLT, using the following argument.

Suppose $X \sim \mathrm{Bi}(n, \theta)$. Then, X is the total number of 'successes' in n independent trials, each with 'success' probability θ. Let X_i be the number of 'successes' in the ith trial ($i = 1, \ldots, n$). Then $X_i = 0$ or 1 with respective probabilities $1 - \theta$ and θ. $E(X_i) = \theta$ and $V(X_i) = \theta(1 - \theta)$.

Now, $X = \sum_{i=1}^{n} X_i$, and $X_1, X_2, \ldots, X_n$ are independent random variables (a property of Bernoulli trials, see Chapter 9). So, applying the CLT, we can justify the following approximation.

Result 17.2

If $X \sim \mathrm{Bi}(n, \theta)$, where $n \geq 20$, $n\theta \geq 5$ and $n(1 - \theta) \geq 5$, then:

$$\frac{X - n\theta}{\sqrt{n\theta(1 - \theta)}} \approx \mathrm{N}(0, 1)$$

Example 3 (continued)

$X \sim \mathrm{Bi}(10{,}000, 0.98)$. So

$$Z = \frac{X - 9800}{14} \approx \mathrm{N}(0, 1)$$

and

$$\begin{aligned} P(X > 9850) &= P(Z > 50/14) \\ &= P(Z > 3.57) \\ &\approx 1 - \Phi(3.57) \\ &= 0.0002 \quad \text{(Appendix 2)} \end{aligned}$$

Example 4

A medical historian wishes to select and archive a small, representative sample of historical case records stored at a local hospital. The historian decides to archive 3% of all the records, to be chosen at random from the store. How many individual records (N) must a category of records contain for this strategy to give probability at least 0.95 that more than 20 of the records are archived?

SOLUTION

Let X be the number of records in this category that are archived. Then $X \sim \text{Bi}(N, 0.03)$ (at least approximately, using the Binomial approximation to the Hypergeometric distribution). So,

$$Z = \frac{X - 0.03N}{\sqrt{0.0297N}} \approx N(0, 1)$$

We require to find N such that

$$P(X > 20) \geq 0.95$$

that is,

$$P\left(Z > \frac{20 - 0.03N}{\sqrt{0.0297N}}\right) \geq 0.95$$

$$1 - \Phi\left(\frac{20 - 0.03N}{\sqrt{0.0297N}}\right) \geq 0.95$$

$$\Phi\left(\frac{20 - 0.03N}{\sqrt{0.0297N}}\right) \leq 0.05 = \Phi(-1.65) \quad \text{(Appendix 2)}$$

$$\frac{20 - 0.03N}{\sqrt{0.0297N}} \leq -1.65$$

$$0.03N - 1.65\sqrt{0.0297N} - 20 \geq 0$$

$$N - 9.48\sqrt{N} - 666.67 \geq 0$$

This quadratic expression in $\sqrt{N}$ has roots at

$$\frac{9.48 \pm \sqrt{9.48^2 + (4 \times 666.67)}}{2}$$

We require $\sqrt{N}$ to be greater than the positive root, that is $\sqrt{N} > 30.99$ or $N \geq 961$.

Example 5

The number of α-particles emitted by a radioactive source in one minute is a Po(36) random variable. What is the probability that the source emits more than 50 particles in one minute? Again, we can justify a Normal approximation to the Poisson distribution that allows us to find an approximate answer to this question.

Suppose $X \sim \text{Po}(\theta)$. Then, X is the total number of 'events' in a certain time interval, say t seconds. Consider dividing the time interval into n non-overlapping segments, each of length t/n seconds. Let X_i be the number of 'events' in the ith time segment ($i = 1, \ldots, n$). Then, assuming a Poisson process (see Chapter 10), $X_i \sim \text{Po}(\theta/n)$, and so $E(X_i) = \theta/n$ and $V(X_i) = \theta/n$.

Now, $X = \sum_{i=1}^{n} X_i$, and $X_1, X_2, \ldots, X_n$ are independent random variables (a property of a Poisson process, see Chapter 10). So, applying the CLT, we can justify the following approximation.

Result 17.3

If $X \sim \text{Po}(\theta)$, where $\theta \geq 30$, then

$$\frac{X - \theta}{\sqrt{\theta}} \approx \text{N}(0, 1)$$

Example 5 (continued)

The number of particles emitted in one minute, X, is a Po(36) random variable. So,

$$Z = \frac{X - 36}{6} \approx \text{N}(0, 1)$$

and

$$P(X > 50) = P(Z > 14/6) \approx 1 - \Phi(14/6) = 1 - \Phi(2.33) = 0.0099$$

(Appendix 2)

EXERCISES ON 17.2

1. Use the Central Limit Theorem to find the following probabilities:
   (a) $P(W > 60)$, when $W \sim \text{Bi}(100, 0.5)$;
   (b) $P(30 < X < 39)$, when $X \sim \text{Bi}(48, 0.75)$;
   (c) $P(Y \leq 20)$, when $Y \sim \text{Po}(30)$;
   (d) $P(Z_1 + Z_2 \leq 40)$, when Z_1, Z_2 i.i.d. Po(30).
2. (a) A complex electronic system consists of 100 components, which all function independently. The probability that a component fails during a normal period of operation is 0.10. The system fails unless at least 90 components continue to function. Find the probability that the system fails during a normal period of operation.
   (b) The system may be modified so that it contains a larger number, N, of components. It will still continue to function as long as at least 90 components function. What value of N is required to reduce the probability of a system failure to 0.05?
3. A double-glazing firm employs 70 sales staff who call on potential customers in the evening. Whenever a sale is concluded, a member of the sales team must phone head office from the customer's home to register the order. The number of orders a sales person achieves in an evening is a Po(0.8) random variable. What is the probability that the head office receives at least 70 calls on an evening when all its sales staff are calling on customers?

Application: acceptance lot sampling

Every manufacturing process produces a proportion of defective items. When the manufactured items are bought in bulk, perhaps for use as parts in a further assembly process, the customer will not want to receive consignments (or lots) that contain a large number of defectives. It is not in the producers' long-term interest to deliver a great many defective items, since this will eventually lose a customer, but the producer will typically expect the customer to be willing to accept a small number of defectives.

Often, the producer and customer agree on a quality assurance procedure. In the simplest procedure, the customer chooses a few items from the lot when it is delivered and tests or inspects them rigorously. If more than an agreed number of them (say, c) are defective, then the whole lot is rejected; otherwise, the whole lot is accepted. This is called **a single-sampling plan**. However good the manufacturing process, there is some probability that a lot will be rejected; however bad the process, there is some probability that a lot will be accepted.

To illustrate this, suppose that a firm produces monitors for computers. They ship their monitors in lots of 1000 items. The customer decides to test 50 monitors from each lot, which represents one day's work for two employees. The customer is willing to accept up to about 5% defectives, and sets $c = 3$ in a single-sampling plan. In other words, the customer will reject the lot if more than three monitors out of 50 tested are found to be defective.

Suppose that M items (out of 1000) are actually defective. Let the random variable X be the number of defective items in the tested sample. Then the probability that a lot is accepted is found from the Hypergeometric distribution of X:

$$P(X \leq 3) = \sum_{x=0}^{3} \binom{M}{x} \binom{1000 - M}{50 - x} \Big/ \binom{1000}{50}$$

This is a difficult probability to calculate exactly, even on a computer. However, the number of items tested (50) is very much smaller than the total lot size (1000), and the proportion defective (hopefully, about 5%) is fairly small. So, we can use a Binomial approximation to the Hypergeometric to find an approximate answer:

$$X \approx \text{Bi}\left(50, \frac{M}{1000}\right)$$

So,

$$P(X \leq 3) \approx \sum_{x=0}^{3} \binom{50}{x} \left(\frac{M}{1000}\right)^{x} \left(1 - \frac{M}{1000}\right)^{50-x}$$

Probabilities of this form can be calculated with less difficulty. However, it is even easier to calculate the probability from a further, Normal, approximation:

$$X \approx N\left(50 \cdot \frac{M}{1000}, 50 \cdot \frac{M}{1000} \cdot \left(1 - \frac{M}{1000}\right)\right)$$

that is,

$$X \approx N\left(50 \cdot \frac{M}{1000}, \frac{50 \cdot M \cdot (1000 - M)}{1000^2}\right)$$

So,

$$P(X \leq 3) \approx \Phi\left(\frac{1000[3 - 50(M/1000)]}{\sqrt{50 \cdot M \cdot (1000 - M)}}\right) = \Phi\left(\frac{3000 - 50M}{\sqrt{50 \cdot M \cdot (1000 - M)}}\right)$$

When $M = 50$, that is 5% of the items are defective, then the probability of accepting the lot is

$$\Phi\left(\frac{500}{\sqrt{50 \cdot 50 \cdot 950}}\right) = \Phi(0.32) = 0.6255 \quad \text{(Appendix 2)}$$

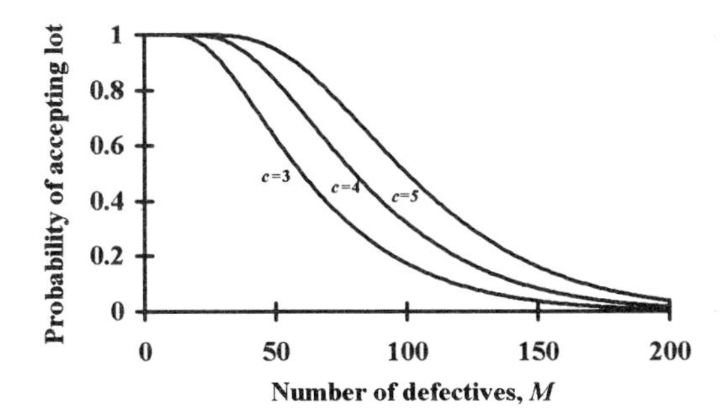

Fig 17.2 Operating characteristic curves for three single-sampling plans.

This might seem a little small, and we might think of increasing c to (say) 4 or 5. Notice, though, that with $c = 3$, the probability of accepting the lot when $M = 60$ is

$$\Phi\left(\frac{0}{\sqrt{50 \cdot 60 \cdot 940}}\right) = \Phi(0) = 0.5000 \quad \text{(Appendix 2)}$$

This might be considered unacceptably high.

An **operating characteristic (OC) curve** is a plot of P(accept the lot) against M. OC curves are plotted in Fig 17.2 for single-sampling plans in this example in which $c = 3$, 4 and 5. In each case, though the probability of accepting the lot is very high for low values of M and almost zero for high values, the probability is not as sensitive as we might like for M between 50 and 100 (i.e. proportions defective between 5% and 10%).

This is a feature of single-sampling plans generally. An alternative, more flexible, procedure is the **double-sampling plan**, which works as follows:

(1) A random sample of n items is drawn from the lot: if at most c of them are defective, then the lot is accepted; if at least d $(> c)$ of them are defective, then the lot is rejected; otherwise, proceed to stage (2).
(2) A second sample of m items is drawn: if at most e of them are defective, then the lot is accepted; otherwise, the lot is rejected.

There are many other variations on the theme of acceptance lot sampling.

Summary

This chapter has shown how an approximate answer may be worked out for the probability of an event associated with the sum or average of a sequence of random variables. The crucial result is the Central Limit Theorem, which justifies approximations based on the Normal distribution. This has been applied, in particular, to give a Normal approximation to the Binomial and the Poisson distributions.

TUTORIAL PROBLEM 1

Choosing a 'random' starting point in the table of random digits in Appendix 1, read off 100 consecutive digits. Display these data in a histogram. Repeat this exercise, but this time read off 100 *pairs* of consecutive digits, average each pair and then display the data. Repeat this exercise a third time, but now averaging *triples* of digits. You should find that the shape of the histogram becomes more like the shape of a Normal probability density function as more digits are averaged. This is a simple demonstration of the Central Limit Theorem.

FURTHER EXERCISES

1. A quadratic detector is a component in a naval sonar system. Let X denote the (scaled) noise level registered on this detector in a 7.5-second interval, when only background noise is present. Then X is a random variable with $E(X) = 2$ and $V(X) = 4$. The system stores and sums the levels registered by the detector in 64 non-overlapping, and hence independent, time intervals. The system automatically signals a possible contact with hostile shipping if this total noise level exceeds a threshold value, d. There is a false alarm if the system signals a contact when only background noise is present. Find the value of d (approximately) that makes the probability of a false alarm 0.05.
2. In a multiple choice examination, each question is accompanied by five possible answers, only one of which is correct. Candidates must choose one answer from the five.
   (a) Suppose that a particular student has probability 0.5 of knowing the correct answer to a question. When the student does not know the correct answer, he or she chooses one of the available answers at random. Show that the chance of answering a question correctly is 0.6.
   (b) Now suppose that the exam consists of 100 questions. A student who gives the correct answer to a question is awarded one mark, but a student who gives the wrong (or no) answer has a quarter of a mark deducted. Write down algebraically the relationship between the number of questions a student answers correctly and the mark obtained for the paper. Assume that a student has probability 0.5 of knowing the answer to every question independently of all the others. Find the probability that the student is awarded more than 50 marks for the paper.
3. A fair coin is tossed $100N$ times (where N is a positive integer), and X is the number of heads obtained. Find the approximate probability that the proportion of heads ($= X/100N$) lies between 0.49 and 0.51. Evaluate this expression for $N = 100$ (10,000 tosses), and find its limiting value as $N \to \infty$.
4. About 10% of bookings on scheduled flights are cancelled too late for airlines to reallocate the seats to other potential passengers. As a result, airlines slightly overbook flights. They hope that, after late cancellations, the number

of passengers checking in for a flight will not exceed the number of available seats.

Suppose that 200 seats are available on a particular flight. Determine how many tickets the airline may issue in advance, while still ensuring that, with probability at least 0.95, 200 or fewer passengers book in for the flight.

5. Each day on which a particular railway station is manned, the number of items handed in to the Lost Property Office is a Po(3) random variable. There are 300 working days in a year (excluding Sundays and Public Holidays). Find the probability that more than 1000 items will be handed in to the office in the course of one year.

18 • Final Thoughts

This book has introduced some fundamental ideas in probability. Starting with the basic concept of a *stochastic experiment*, Chapters 1 to 7 built up results required to calculate the probabilities of *events* in a *sample space*. These were primarily consequences of the *Axioms of Probability* and the definition of *conditional probability*. In Chapters 8 to 14, the focus switched to properties of *random variables*, particularly their *probability distribution* or *probability density function*, their *expected value* and *variance*, and (for more theoretical purposes) their *moment-generating function*. Chapter 15 began to consider situations in which more than one random variable is required to describe the outcome of an experiment adequately. This led naturally to an interest in the sum or average of a sequence of random variables, culminating in the *Central Limit Theorem*, one of the most remarkable and useful theorems in mathematics.

This book has been written to highlight the practical applications of probability. The Applications sections have introduced particular uses of probability in physics, genetics, medicine, engineering and sport. Some of the examples have touched on historical, sociological and forensic applications. The purpose of this effort has been to suggest that probability, although of mathematical interest in itself, is worth studying because of its usefulness in solving real-life problems of considerable consequence.

There is no shortage of topics that could be added to the material covered in this book. At the simplest level, there are important distributions to discuss other than the most common ones described so far, proofs that should be added or made more rigorous, results that could be generalized and more advanced problems that could be tackled using the concepts already introduced.

Completely new material could be developed from the concept of a bivariate distribution, which has been dealt with very sketchily in this book. For example, nothing much has been said about the joint probability density function of continuous random variables. Consideration needs to be given to methods for deriving the joint distribution of two functions of two random variables. These results for two random variables need to be generalized further, to the case of an arbitrary (finite) number of random variables (**multivariate probability**). All of this work finds important uses in **statistical inference**, which is itself a tool for drawing valid conclusions from numerical data in many different disciplines.

When discussing the Poisson process (Chapter 10), we touched on another advanced topic in probability, **stochastic processes**. A stochastic process describes how a system that is subject to uncertainty evolves through time. An important application is in the modelling of populations, such as a human population subject to the effects of birth, death and immigration, or an ecological system typified by the interaction of two species (predator–prey models). Epidemics are modelled by similar methods; for example, at present there is considerable epidemiological interest in the world-wide AIDS epidemic. Some engineering systems, too, are modelled by stochastic processes, for example the levels of water in reservoirs or behind hydroelectric dams. On a lighter note, the score in a game of squash could be described by a stochastic process.

The mention of biological and engineering systems brings to mind the topic of **simulation**. Here, the probabilistic behaviour of a complex system is mimicked on a computer. Usually, the motivation for simulating a system is that its complexity rules out an analytical solution to real problems. The basic building block of a simulation program is an algorithm that generates realizations of individual random variables. For example, the reliability of the fire-warning system on a ship could be simulated by a computer program that generates a value for the exponentially distributed lifetime of every individual component.

Getting away from these essentially applied aspects of probability, the theoretical foundations of the subject have been explored systematically in the mathematical subject of **measure theory**.

In all these different ways, the subject of probability develops from the foundations laid in this book.

Appendix I: Table of Random Digits

To use this table, start at an arbitrary point. Read the required number of digits consecutively, across rows, from left to right.

94768	85505	05722	34619	19396	95253	05390	21007	61304	13481
52738	62275	53292	84989	84566	67223	56160	02577	05674	32651
38708	87045	68862	23359	57736	67193	54930	52147	38044	59821
52678	59815	52432	49729	38906	95163	01700	69717	58414	94991
94648	80585	04002	64099	28076	51133	96470	55287	66784	38161
64618	49355	23572	66469	25246	35103	39240	08857	63154	89331
62588	66125	11142	56839	30416	47073	30010	30427	47524	48501
88558	30895	66712	35209	43586	87043	68780	19997	19894	15671
42528	43665	90282	01579	64756	55013	55550	77567	80264	90841
24498	04435	81852	55949	93926	50983	90320	03137	28634	74011
34468	13205	41422	98319	31096	74953	73090	96707	65004	65181
72438	69975	68992	28689	76266	26923	03860	58277	89374	64351
38408	74745	64562	47059	29436	06893	82630	87847	01744	71521
32378	27515	28132	53429	90606	14863	09400	85417	02114	86691
54348	28285	59702	47799	59776	50833	84170	50987	90484	09861
04318	77055	59272	30169	36946	14803	06940	84557	66854	41031
82288	73825	26842	00539	22116	06773	77710	86127	31224	80201
88258	18595	62412	58909	15286	26743	96480	55697	83594	27371
22228	11365	65982	05279	16456	74713	63250	93267	23964	82541
84198	52135	37552	39649	25626	50683	78020	98837	52334	45711
74168	40905	77122	62019	42796	54653	40790	72407	68704	16881
92138	77675	84692	72389	67966	86623	51560	13977	73074	96051
38108	62445	60262	70759	01136	46593	10330	23547	65444	83221
12078	95215	03832	57129	42306	34563	17100	01117	45814	78391
14048	75985	15402	31499	91476	50533	71870	46687	14184	81561
44018	04755	94972	93869	48646	94503	74640	60257	70554	92731
01988	81525	42542	44239	13816	66473	25410	41827	14924	11901
87958	06295	58112	82609	86986	66443	24180	91397	47294	39071
01928	79065	41682	08979	68156	94413	70950	08967	67664	74241
43898	99835	93252	23349	57326	50383	65720	94537	76034	17411

13868	68605	12822	25719	54496	34353	08490	48107	72404	68581
11838	85375	00392	16089	59666	46323	99260	69677	56774	27751
37808	50145	55962	94459	72836	86293	38030	59247	29144	94921
91778	62915	79532	60829	94006	54263	24800	16817	89514	70091
73748	23685	71102	15199	23176	50233	59570	42387	37884	53261
83718	32455	30672	57569	60346	74203	42340	35957	74254	44431
21688	89225	58242	87939	05516	26173	73110	97527	98624	43601
87658	93995	53812	06309	58686	06143	51880	27097	10994	50771
81628	46765	17382	12679	19856	14113	78650	24667	11364	65941
03598	47535	48952	07049	89026	50083	53420	90237	99734	89111

82218	70955	09172	76069	18846	72703	80840	14457	92754	02931
20188	27725	36742	06439	64016	24673	11610	76027	17124	02101
86158	32495	32312	24809	17186	04643	90380	05597	29494	09271
80128	85265	95882	31179	78356	12613	17150	03167	29864	24441
02098	86035	27452	25549	47526	48583	91920	68737	18234	47611
53568	96305	48522	89419	66196	14053	76190	23807	76104	20281
31538	93075	16092	59789	51366	06023	46960	25377	40474	59451
37508	37845	51662	18159	44536	25993	65730	94957	92844	06621
71478	30615	55232	64529	45706	73963	32500	32517	33214	61791
33448	71385	26802	98899	54876	49933	47270	38087	61584	24961

23418	60155	66372	21269	72046	53903	10040	11657	77954	96131
41388	96925	73942	31639	97216	85873	20810	53227	82324	75301
87358	81695	49512	30009	30386	45843	79580	62797	74694	62471
61328	14465	93082	16379	71566	33813	86350	40367	55064	57641
63298	95235	04652	90749	20726	49783	41120	85937	23434	60811
93268	24005	84222	53119	77896	93753	43890	99507	79804	71981
51238	00775	31792	03489	43066	65723	94660	81077	24174	91151
37208	25545	47362	41859	16236	65693	93430	30647	56544	18321
51178	98315	30932	68229	97406	93663	40200	48217	76914	53491
93148	19085	82502	82599	86576	49633	34970	33787	85284	96661

63118	87855	02072	84969	83746	33603	77740	87357	81654	47831
61088	04625	89642	75339	88916	45573	68510	08927	66024	07001
87058	69395	45212	53709	02086	85543	07280	98497	38394	74171
41028	82165	68782	20079	23256	53513	94050	56067	98764	49341
22998	42935	60352	74449	52426	49483	28820	81637	47134	32511
32968	51705	19922	16819	89596	73453	11590	75207	83504	23681
70938	08475	47492	47189	34766	25423	42360	36777	07874	22851
36908	13245	43062	65559	87936	05393	21130	66347	20244	30021
30878	66015	06632	71929	49106	13363	47900	63917	20614	45191
52848	66785	38202	66299	18276	49333	22670	29487	08984	68361

02818	15555	37772	48669	95446	13303	45440	63057	85354	99531
80788	12325	05342	19039	80616	05273	16210	64627	49724	38701
86758	57095	40912	77409	73786	25243	34980	34197	02094	85871
20728	49865	44482	23779	74956	73213	01750	71767	42464	41041

Appendix 2: Table of the Standard Normal Cumulative Distribution Function

When Z is an N(0, 1) random variable, this table gives $\Phi(z) = P(Z \leq z)$ for values of z from 0.00 to 3.67 in steps of 0.01.

z	0	1	2	3	4	5	6	7	8	9
0.0	0.5000	0.5040	0.5080	0.5120	0.5160	0.5199	0.5239	0.5279	0.5319	0.5359
0.1	0.5398	0.5438	0.5478	0.5517	0.5557	0.5596	0.5636	0.5675	0.5714	0.5753
0.2	0.5793	0.5832	0.5871	0.5910	0.5948	0.5987	0.6026	0.6064	0.6103	0.6141
0.3	0.6179	0.6217	0.6255	0.6293	0.6331	0.6368	0.6406	0.6443	0.6480	0.6517
0.4	0.6554	0.6591	0.6628	0.6664	0.6700	0.6736	0.6772	0.6808	0.6844	0.6879
0.5	0.6915	0.6950	0.6985	0.7019	0.7054	0.7088	0.7123	0.7157	0.7190	0.7224
0.6	0.7257	0.7291	0.7324	0.7357	0.7389	0.7422	0.7454	0.7486	0.7517	0.7549
0.7	0.7580	0.7611	0.7642	0.7673	0.7704	0.7734	0.7764	0.7794	0.7823	0.7852
0.8	0.7881	0.7910	0.7939	0.7967	0.7995	0.8023	0.8051	0.8078	0.8106	0.8133
0.9	0.8159	0.8186	0.8212	0.8238	0.8264	0.8289	0.8315	0.8340	0.8365	0.8389
1.0	0.8413	0.8438	0.8461	0.8485	0.8508	0.8531	0.8554	0.8577	0.8599	0.8621
1.1	0.8643	0.8665	0.8686	0.8708	0.8729	0.8749	0.8770	0.8790	0.8810	0.8830
1.2	0.8849	0.8869	0.8888	0.8907	0.8925	0.8944	0.8962	0.8980	0.8997	0.9015
1.3	0.9032	0.9049	0.9066	0.9082	0.9099	0.9115	0.9131	0.9147	0.9162	0.9177
1.4	0.9192	0.9207	0.9222	0.9236	0.9251	0.9265	0.9279	0.9292	0.9306	0.9319
1.5	0.9332	0.9345	0.9357	0.9370	0.9382	0.9394	0.9406	0.9418	0.9429	0.9441
1.6	0.9452	0.9463	0.9474	0.9484	0.9495	0.9505	0.9515	0.9525	0.9535	0.9545
1.7	0.9554	0.9564	0.9573	0.9582	0.9591	0.9599	0.9608	0.9616	0.9625	0.9633
1.8	0.9641	0.9649	0.9656	0.9664	0.9671	0.9678	0.9686	0.9693	0.9699	0.9706
1.9	0.9713	0.9719	0.9726	0.9732	0.9738	0.9744	0.9750	0.9756	0.9761	0.9767
2.0	0.9772	0.9778	0.9783	0.9788	0.9793	0.9798	0.9803	0.9808	0.9812	0.9817
2.1	0.9821	0.9826	0.9830	0.9834	0.9838	0.9842	0.9846	0.9850	0.9854	0.9857
2.2	0.9861	0.9864	0.9868	0.9871	0.9875	0.9878	0.9881	0.9884	0.9887	0.9890
2.3	0.9893	0.9896	0.9898	0.9901	0.9904	0.9906	0.9909	0.9911	0.9913	0.9916
2.4	0.9918	0.9920	0.9922	0.9925	0.9927	0.9929	0.9931	0.9932	0.9934	0.9936
2.5	0.9938	0.9940	0.9941	0.9943	0.9945	0.9946	0.9948	0.9949	0.9951	0.9952
2.6	0.9953	0.9955	0.9956	0.9957	0.9959	0.9960	0.9961	0.9962	0.9963	0.9964
2.7	0.9965	0.9966	0.9967	0.9968	0.9969	0.9970	0.9971	0.9972	0.9973	0.9974
2.8	0.9974	0.9975	0.9976	0.9977	0.9977	0.9978	0.9979	0.9979	0.9980	0.9981
2.9	0.9981	0.9982	0.9983	0.9983	0.9984	0.9984	0.9985	0.9985	0.9986	0.9986

z	0	1	2	3	4	5	6	7	8	9
3.0	0.9987	0.9987	0.9987	0.9988	0.9988	0.9989	0.9989	0.9989	0.9990	0.9990
3.1	0.9990	0.9991	0.9991	0.9991	0.9992	0.9992	0.9992	0.9992	0.9993	0.9993
3.2	0.9993	0.9993	0.9994	0.9994	0.9994	0.9994	0.9994	0.9995	0.9995	0.9995
3.3	0.9995	0.9995	0.9996	0.9996	0.9996	0.9996	0.9996	0.9996	0.9996	0.9997
3.4	0.9997	0.9997	0.9997	0.9997	0.9997	0.9997	0.9997	0.9997	0.9997	0.9998
3.5	0.9998	0.9998	0.9998	0.9998	0.9998	0.9998	0.9998	0.9998	0.9998	0.9998
3.6	0.9998	0.9998	0.9998	0.9998	0.9998	0.9998	0.9998	0.9999		

When $z < 0.00$, $\Phi(z)$ can be found from this table using the relationship $\Phi(-z) = 1 - \Phi(z)$.

Answers to Selected Exercises

Answers are provided only for the exercises within chapters, not for the Further Exercises at the ends of chapters.

Section 2.1

1. (a) $\{0, 1, 2, 3\}$; (b) $\{0, 1, 2, \ldots\}$; (c) $\{x : x > 0\}$;
   (d) $\{x : x \in \mathbf{R}\}$, where $x =$ husband's height $-$ wife's height;
   (e) $\{x : x \geq 0\}$; (f) $\{(x, y) : x \geq 0; y > 0\}$; (g) $\{0, 1, 2, \ldots, 100\}$;
   (h) $\{(x, y) : x = 0, 1, 2, \ldots, 100; y = 0, 1, 2, \ldots, 100\}$.

Section 2.2

1. (a) $A = \{2, 3\}$; $B = \{0, 1\}$; (b) $E = \{0, 1, 2\}$;
   (c) $F = \{x : 0 < x < 10\}$; $G = \{x : x > 5\}$;
   (d) $A = \{x : x < 0\}$; (f) $C = \{(x, y) : x > y > 0\}$;
   (h) $D = \{(x, y) : x = 75, 76, \ldots, 100; y = 0, 1, \ldots, 100\}$;
   $E = \{(x, y) : x = 0, 1, \ldots, 100; y = 75, 76, \ldots, 100\}$;
   $F = \{(x, y) : x = 0, 1, \ldots, 100; y = 0, 1, \ldots, 100 \text{ and } x + y \geq 150\}$.
2. (a) $S_1 = \{\text{HHH, HHT, HTH, HTT, THH, THT, TTH, TTT}\}$;
   (b) $S_2 = \{0, 1, 2, 3\}$; (c) $S_3 = \{0, 1, 2, 3\}$;
   A can be represented in S_1 and S_3;
   B can be represented in S_1 and S_3;
   C can be represented in S_1 and S_2;
   D can be represented in S_1.

Section 2.3

1. (a) $A \cup B = \{0, 1, 2, 3\}$, $A \cap B = \emptyset$, $A' = \{0, 1\} = B$;
   (b) $E' = \{3, 4, 5, \ldots\} =$ 'you will visit the doctor at least three times in the next year';
   (c) $F \cap G = \{x : 5 < x < 10\}$, $F \cup G = \{x : x > 0\}$;
   (h) $D \cap E = \{(x, y) : x = 75, 76, \ldots, 100; y = 75, 76, \ldots, 100\}$,
   $D \cap E \cap F = D \cap E$.

Section 3.1

1. The (cumulative) relative frequencies are:
   0.300, 0.275, 0.250, 0.250, 0.250, 0.283, 0.264, 0.256,
   0.256, 0.255, 0.250, 0.258, 0.258, 0.261, 0.263, 0.263.
2. The (cumulative) relative frequencies are:
   0.500, 0.475, 0.483, 0.463, 0.450, 0.458, 0.457, 0.444,
   0.439, 0.440, 0.441, 0.442, 0.458, 0.461, 0.467, 0.463.

Section 4.1

1. (1, 3, 4) 6 (2, 3, 4) 5 (3, 3, 4) 4
(1, 3, 5) 7 (2, 3, 5) 6 (3, 3, 5) 5
(1, 3, 6) 8 (2, 3, 6) 7 (3, 3, 6) 6
(1, 4, 4) 7 (2, 4, 4) 6 (3, 4, 4) 5
(1, 4, 5) 8 (2, 4, 5) 7 (3, 4, 5) 6
(1, 4, 6) 9 (2, 4, 6) 8 (3, 4, 6) 7
(1, 5, 4) 8 (2, 5, 4) 7 (3, 5, 4) 6
(1, 5, 5) 9 (2, 5, 5) 8 (3, 5, 5) 7
(1, 5, 6) 10 (2, 5, 6) 9 (3, 5, 6) 8
Of the 27 outcomes, 14 give even scores. So $P(\text{score is even}) = 14/27 = 0.519$.
2. For $i = 1, 2, \ldots, k$, define the following procedure:
η_i: decide whether or not to include the ith element in the subset (two choices)
By MP, the total number of possible subsets is $2 \times 2 \times \ldots \times 2 = 2^k$. (Notice that the subset with no elements is the empty set.)
3. 0.1
(a) 0.01; (b) 0.72; (c) 0.27.

Section 4.2

1. (a) $4! = 4 \times 3 \times 2 \times 1 = 24$; (b) $7! = 7 \times 6 \times 5 \times 4 \times 3 \times 2 \times 1 = 5040$;
(c) $\frac{7!}{4!} = \frac{7 \cdot 6 \cdot 5 \cdot 4 \cdot 3 \cdot 2 \cdot 1}{4 \cdot 3 \cdot 2 \cdot 1} = 7 \times 6 \times 5 = 210$;
(d) $\frac{6!}{5!} = \frac{6 \cdot 5 \cdot 4 \cdot 3 \cdot 2 \cdot 1}{5 \cdot 4 \cdot 3 \cdot 2 \cdot 1} = 6$; (e) $\frac{3!}{0!} = \frac{3 \cdot 2 \cdot 1}{1} = 6$.
2. (a) $n = 5, r = 5$; 120 permutations;
(b) $n = 6, r = 0$; one permutation;
(c) $n = 8, r = 2$; 56 permutations.
3. (a) $n = 7, n_1 = n_2 = n_3 = n_4 = n_5 = n_6 = n_7 = 1$; 5040 permutations;
(b) $n = 7, n_1 = 3, n_2 = n_3 = n_4 = n_5 = 1$; 840 permutations;
(c) $n = 7, n_1 = 3, n_2 = 2, n_3 = n_4 = 1$; 420 permutations.
4. There are $8! = 40{,}320$ possible arrangements altogether. The suspect is either first or last in $2 \times 7! = 10{,}080$ of these arrangements. Assuming that all the arrangements are equally likely, then the probability that the suspect is either first or last is 0.25.
5. $n = 8, n_1 = n_2 = n_3 = n_4 = 2$; 2520 permutations.

Section 4.3

1. (a) $n = 7, r = 2$; 21 combinations; (b) $n = 7, r = 5$; 21 combinations;
(c) $n = 7, r = 3$; 35 combinations; (d) $n = 8; r = 8$; one combination.

2.

r	0	1	2	3	4	5	6
$\binom{6}{r}$	1	6	15	20	15	6	1

3. Number of different ways to choose five athletes to run in heat A is:

$$\binom{10}{5} = \frac{10 \cdot 9 \cdot 8 \cdot 7 \cdot 6}{1 \cdot 2 \cdot 3 \cdot 4 \cdot 5} = 252.$$

(a) 0.083; (b) 0.167; (c) 0.834.

4. (a) 210; (b) $100/210 = 0.476$; (c) $70/210 = 0.333$.

5. Number of dominos $= 7 + \binom{7}{2} = 7 + 21 = 28$.

6. (a) Binomial Theorem with $x = y = 1$.
(b) Binomial Theorem with $x = 1 - \theta$ and $y = \theta$.

Section 5.1

1. Axioms 1 and 2 follow immediately from the definition of the probability of an event in an equally likely outcomes model. Axiom 3 is just the Addition Principle. Exercise 2 shows that a finite sample space can have only a finite number of different events, so Axiom 4 is not required in this case.

Section 5.2

1. $P(A \cup B) = 1 - 0.45 = 0.55$;
$P(A \cap B) = P(A) + P(B) - P(A \cup B) = 0.30 + 0.35 - 0.55 = 0.10$.

3. Suppose that $S = \{e_1, e_2, \ldots, e_k\}$. For $i = 1, 2, \ldots, k$, let the event $E_i = \{e_i\}$. Then, $S = E_1 \cup E_2 \cup \ldots \cup E_k$. By Axiom 2, $P(E_1 \cup E_2 \cup \ldots \cup E_k) = P(S) = 1$. But, $E_1, E_2, \ldots, E_k$ are all disjoint. By Axiom 4, then, it follows that $P(E_1) + P(E_2) + \ldots + P(E_k) = P(E_1 \cup E_2 \cup \ldots \cup E_k) = 1$.

Section 6.1

1.

Age group	$P(\text{dies} \mid \text{male})$	$P(\text{dies} \mid \text{female})$	$P(\text{dies})$
65–74	0.045	0.027	0.035
75–84	0.101	0.065	0.078
85+	0.226	0.170	0.183

2. (a) $19/20 = 0.95$; (b) $18/19 = 0.947$; (c) 1.

Section 6.2

1. (a) 0.4; (b) 0.162; (c) 0.508.

2. 0.882.

3. 0.66.

Section 6.3

2. $P(\text{both engines fail}) = \theta^2 < \theta$.

3. $P(A) = P(B) = P(C) = 0.5$;
$P(A \cap B) = P(A \cap C) = P(B \cap C) = 0.25$, so the events are pairwise independent;
$P(A \cap B \cap C) = 0 \neq P(A) \cdot P(B) \cdot P(C)$, so the events are not independent.

4. $P(A) = P(B) = P(C) = 0.5$;
$P(A \cap B \cap C) = P(\{3\}) = 1/8 = P(A) \cdot P(B) \cdot P(C)$;
$P(A \cap B) = P(\{1, 3, 4\}) = 3/8 \neq P(A) \cdot P(B)$;
$P(A \cap C) = P(\{3\}) = 1/8 \neq P(A) \cdot P(C)$;
$P(B \cap C) = P(\{3\}) = 1/8 \neq P(B) \cdot P(C)$;
so the events are not independent.

Section 7.1

1. 0.433 (UK); 0.5 (West of Scotland).
2. 0.279.

Section 7.2

1. (a) 0.462 (UK); (b) 0.667 (West of Scotland).
2. 0.670.

Section 8.1

1. (a) $\{90, 120, 210, 240\}$; (b) $\{-10, -9, \ldots, 9, 10\}$;
(c) $\{-5, -3\frac{1}{2}, \ldots, 8\frac{1}{2}, 10\}$.
2. Only $p_2(x)$ is a valid probability distribution.

3.

x	2	3	4	5	6	7	8	9	10	11	12
$p(x)$	1/36	2/36	3/36	4/36	5/36	6/36	5/36	4/36	3/36	2/36	1/36

4.

x	0	1	2
$p(x)$	0.83	0.09	0.08

Section 9.1

1.

x	0	1	2	3	4
$p(x)$	0.316	0.422	0.211	0.047	0.004
$F(x)$	0.316	0.738	0.949	0.996	1.000

2.

y	0	1	2	3	4
$p(y)$	0.004	0.047	0.211	0.422	0.316
$F(y)$	0.004	0.051	0.262	0.684	1.000

$p_Y(y) = p_X(4 - y), y = 0, 1, 2, 3, 4.$
$F_Y(y) = 1 - F_X(3 - y), y = 0, 1, 2, 3.$

4. $0.794, p(x) = \left[\binom{20}{x}\right](0.2)^x(0.8)^{20-x}$.
5. 0.0313.

Section 9.2

1. 0.8, 0.16, 0.04, 0.992.
2. $F(x) = 1 - \theta^x, x = 1, 2, \ldots$.
3. (a) 0.15; (b) 0.0225.

Section 10.1

1.

x	0	1	2	3	4	5	...
$p(x)$	0.6065	0.3033	0.0758	0.0126	0.0016	0.0002	...
$F(x)$	0.6065	0.9098	0.9856	0.9982	0.9998	1.0000	...

2. (a) 0.632; (b) $Y \sim \text{Ge}(0.632)$.
3. 0.1088.
4. 0.001.

Section 10.2

1.

x	0	1	2	3
$p(x)$	0.2637	0.4945	0.2198	0.0220

2. (a) Hyp(20, 200, 2); prob = 0.1905; (b) Bi(20, 0.01); prob $\cong$ 0.1821.
3. 0.4387.
4. 0.5780.

Section 11.1

1. $E(X) = 0$; $E(X) = -0.25$; $E(X) = 0.25$.

2.

x	1	2	3
$p(x)$	0.01	0.27	0.72

$E(X) = 2.71$.

Section 11.2

2. (a) $E(X) = 0, E(X^2) = 1, V(X) = 1$.
   (b) $E(X) = 2\theta - 1, E(X^2) = 1, V(X) = 4\theta(1 - \theta)$.
3. $V(X) = 0.5$; $V(X) = 0.6875$; $V(X) = 0.6875$.
4. (a) $E(X) = 10, V(X) = 9, \text{sd}(X) = 3$;
   (b) $E(X) = 20, V(X) = 16, \text{sd}(X) = 4$;
   (c) $E(X) = 50, V(X) = 25, \text{sd}(X) = 5$;
   (d) $E(X) = 90, V(X) = 9, \text{sd}(X) = 3$.

Section 12.1

1. 0, 0.25, 0.25, 0.32, 0.36.
2. $\frac{1}{\sqrt{2}}, \frac{\sqrt{3} - 1}{2}, 1 - \frac{\sqrt{3}}{2}$.

Section 12.2

1. $f(x) = 2x (0 < x < 1); f(x) = \cos x (0 < x < \pi/2); f(x) = \theta^2 x e^{-\theta x} \ (x > 0)$.
2. (a) $F(x) = 3x^2 - 2x^3 \ (0 < x < 1)$; (b) $F(x) = 1 - c^3/x^3 \ (x > c)$;
   (c) $F(x) = 1 - \exp(-x^2) \ (x > 0)$.
3. $k = 12$; (a) 0.5248; (b) 0.0272.

Section 12.3

2. $E(X) = (a+b)/2$, $V(X) = (b-a)^2/12$.
3. (a) $E(X) = 0.5$, $V(X) = 0.05$; (b) $E(X) = 1.5c$, $V(X) = 0.75c^2$;
   (c) $E(X) = \dfrac{\sqrt{\pi}}{2}$, $V(X) = 1 - \dfrac{\pi}{4}$.
5. $P\left(X > \dfrac{1}{\theta}\right) = \dfrac{1}{e}$.

Section 12.4

1. $E(X) = \dfrac{2}{\theta}$, $V(X) = \dfrac{2}{\theta^2}$.
2. $M_X(u) = \dfrac{e^{bu} - e^{au}}{(b-a)u}$.

Section 13.1

1. (a) $\Phi(1.65) = 0.9505$; (b) $1 - \Phi(-1.65) = \Phi(1.65) = 0.9505$;
   (c) $2\Phi(1.65) - 1 = 0.9010$; (d) $\Phi(1.28) = 0.8997$;
   (e) $\Phi(1.65) - \Phi(1.28) = 0.0508$.
   $c = 2.57$.
2. $Z = \dfrac{X - 20}{4} \sim N(0, 1)$.

   (a) $1 - \Phi(3) = 0.0013$; (b) $2\Phi(1.5) - 1 = 0.8664$;
   (c) $1 - \Phi(2.5) = 0.0062$.
3. $Z = \dfrac{X + 4}{2} \sim N(0, 1)$.

   (a) $\Phi(2) = 0.9772$; (b) $2\Phi(1) - 1 = 0.6826$; (c) $1 - \Phi(1.75) = 0.0401$.
   $c = -7.29$.
4. (a) The probabilities are 0.8413 and 0.5, so the first route is better.
   (b) The probabilities are 0.9772 and 0.9938, so the second route is better.

Section 14.1

1. (a)

y	0	$\frac{1}{4}$	$\frac{1}{2}$	$\frac{3}{4}$	1
$p_Y(y)$	0.09	0.12	0.18	0.36	0.25

(b)

y	0	1	2	3	4
$p_Y(y)$	0.25	0.36	0.18	0.12	0.09

2.

y	0	1	4
$p_Y(y)$	0.4	0.4	0.2

3. $X \sim \text{Hyp}(5, 25, 3)$. $Y = 50$ when $X = 0$; $Y = 100$ when $X = 1$; $Y = 400$ when $X = 2$ or 3. So:

y	50	100	400
$p_Y(y)$	0.4957	0.4130	0.0913

$E(Y) = £110.14$.

4. $X \sim \text{Po}(0.5)$. $Y = 0$ when $X = 0$; $Y = 10$ when $X = 1$; $Y = 20$ otherwise. So:

y	0	10	20
$p_Y(y)$	0.6065	0.3033	0.0902

$E(Y) = £4.84$.

Section 14.2

2. (a) $f_Y(y) = \dfrac{15\sqrt{y}}{8\pi\sqrt{\pi}}\left(1 - \sqrt{\dfrac{y}{4\pi}}\right)^2, 0 < y < 4\pi.$

(b) $f_Z(z) = \dfrac{30(3z)^{4/3}}{(4\pi)^{7/3}}\left[1 - \left(\dfrac{3z}{4\pi}\right)^{1/3}\right]^2, 0 < y < \dfrac{4\pi}{3}.$

5. $f_Y(y) = \dfrac{1}{\pi(1+y)\sqrt{y}}, y > 0.$

Section 15.1

1. (a) 0.6, 0.3, 0.2.

(b)

x	$-\frac{1}{2}$	0	$\frac{1}{2}$	1
$p_X(x)$	0.3	0.1	0.3	0.3

y	-1	$-\frac{1}{2}$	0	$\frac{1}{2}$
$p_Y(y)$	0.3	0.3	0.1	0.3

(c) $E(2X + 2Y) = 0$.

2. (a) 0.34, 0.64, 0.49.

(b)

y	0	1	5
$p_Y(y)$	0.4	0.3	0.3

$P(Y > 0) = 0.6$.

(c) $E(X) = E(Y) = 1.8$; $E(XY) = 3.24$.

3. (a) 0.0077, 0.0146;
   (b) $X \sim \text{Bi}(10, \frac{1}{2})$, $P(X > 7) = 0.055$, $E(X) = 5$, $V(X) = 2.5$;
   (c) $Y \sim \text{Bi}(10, \frac{1}{4})$, $P(Y \leq 3) = 0.7759$, $E(Y) = 2.5$, $V(Y) = 1.875$.

Section 15.2

1. (a)

x	$-\frac{1}{2}$	$\frac{1}{2}$	1
$p_{X\mid Y}(x \mid \frac{1}{2})$	1/3	1/3	1/3

(b)

x	0
$p_{X\mid Y}(x \mid 0)$	1

(c)

y	-1	$-\frac{1}{2}$	$\frac{1}{2}$
$p_{Y\mid X}(y \mid -\frac{1}{2})$	1/3	1/3	1/3

2.

x	0	1	5
$p_X(x)$	0.4	0.3	0.3
$p_{X\mid Y}(x \mid 0)$	0.4	0.3	0.3
$p_{X\mid Y}(x \mid 1)$	0.4	0.3	0.3
$p_{X\mid Y}(x \mid 5)$	0.4	0.3	0.3

3. (a) $\text{Bi}(8, \frac{2}{3})$; (b) $\text{Bi}(2, \frac{1}{2})$.

Section 15.3

1.

z	-3	-2	-1	0	1	2	3
$p_Z(z)$	0.1	0.1	0.1	0.4	0.1	0.1	0.1

2.

z	0	1	5	25
$p_Z(z)$	0.64	0.09	0.18	0.09

Section 16.1

1. (a) 0, 2; (b) 1, 2; (c) 10, 8; (d) 0, 25.
2. Assuming that the numbers of new customers who contact the consultancy on different days are independent, $E(X) = V(X) = 10$.
3. (a) $E(Y) = \theta$; $V(Y) = \dfrac{\theta}{\sum t_i}$; (b) $E(Z) = \theta$; $V(Z) = \dfrac{\theta}{m}\sum \dfrac{1}{t_i}$.

Section 16.2

1. $\bar{X} \sim N(\mu, \sigma^2/m)$; $Y \sim N(m\mu, m\sigma^2)$.
2. $X - Y \sim N(0, 0.0008)$; the probability that a pair is not satisfactory is 0.2892.

Section 17.1

1. 173.7 hours.

Section 17.2

1. (a) $W \approx \mathrm{N}(50, 25)$, $P(W > 60) = 0.0228$;
   (b) $X \approx \mathrm{N}(36, 9)$, $P(30 < X < 39) = 0.8185$;
   (c) $Y \approx \mathrm{N}(30, 30)$, $P(Y \leq 20) = 0.0336$;
   (d) $Z_1 + Z_2 \approx \mathrm{N}(60, 60)$, $P(Z_1 + Z_2 \leq 40) = 0.0049$.
2. (a) 0.5; (b) $N \geq 106$.
3. 0.0307.

Bibliography

Barnett, V. *Comparative Statistical Inference*, 2nd edn, London, John Wiley, 1982.

Crossley, J. A., Aitken, D. A. and Connor, J. M. Prenatal screening for chromosome abnormalities using maternal serum chorionic gonadotrophin, alphafetoprotein, and age. *Prenatal Diagnosis* **11**, 83–101, 1991.

Feller, W. *An Introduction to Probability Theory and its Applications, Volume 1*, 3rd edn, New York, John Wiley, 1968.

Fisher, N., Turner, S. W., Pugh, R. and Taylor, C. Estimating numbers of homeless and homeless mentally ill people in north east Westminster by using capture–recapture analysis. *British Medical Journal* **308**, 27–30, 1994.

George, S. L. Optimal strategy in tennis: a simple probabilistic model. *Applied Statistics* **22**, 97–104, 1973.

Hart, J. T. Rule of halves: implications of increasing diagnosis and reducing dropout for future workload and prescribing costs in primary care. *The British Journal of General Practice* **42**, 116–19, 1992.

Hirst, K. E. *Numbers, Sequences and Series*, London, Edward Arnold, 1995.

Hsi, B. P. and Burych, D. M. Games of two players. *Applied Statistics* **20**, 86–92, 1971.

Kolmogorov, A. *Grundbegriffe der wahrscheinlichkeitsrechnung*, Berlin, Springer, 1933. English edition (translated by N. Morrison) *Foundations of the Theory of Probability*, Chelsea, New York, 1956.

Laporte, R. E. Assessing the human condition: capture–recapture techniques. *British Medical Journal* **308**, 5–6, 1994.

Lindley, D. V. Introduction to probability and statistics from a Bayesian viewpoint, Part 1. *Probability*, Cambridge, Cambridge University Press.

McKeganey, N., Barnard, M., Leyland, A., Coote, I. and Follet, E. Female streetworking prostitutes and HIV infection in Glasgow. *British Medical Journal* **305**, 801–4, 1992.

Murphy, E. A. and Mutalik, G. S. The application of Bayesian methods in genetic counselling. *Human Heredity* **19**, 126–51, 1969.

Reep, C. and Benjamin, B. Skill and chance in association football. *Journal of the Royal Statistical Society, Series A* **131**, 581–5, 1968.

Reep, C., Pollard, R. and Benjamin, B. Skill and chance in ball games. *Journal of the Royal Statistical Society, Series A* **134**, 623–9, 1971.

Registrar General for Scotland, *Annual Report of the Registrar General for Scotland for the year 1992*, London, HMSO, 1993.

Simmons, J. A probabilistic model of squash: strategies and applications. *Applied Statistics* **38**, 95–110, 1989.

Strauss, D. and Arnold, B. C. The rating of players in racquetball tournaments. *Applied Statistics* **36**, 163–73, 1987.

Woodward, L. A. *Molecular Statistics for Students of Chemistry*, Oxford, Clarendon Press, 1975.

Zweig, M. H. and Campbell, G. Receiver operating characteristic (ROC) plots: a fundamental evaluation tool in clinical medicine. *Clinical Chemistry* **39**, 561–77, 1993.

Index